Time, Technology and Environment

Plateaus – New Directions in Deleuze Studies

'It's not a matter of bringing all sorts of things together under a single concept but rather of relating each concept to variables that explain its mutations.'
Gilles Deleuze, *Negotiations*

Series Editors

Ian Buchanan, University of Wollongong
Claire Colebrook, Penn State University

Editorial Advisory Board

Keith Ansell Pearson
Ronald Bogue
Constantin V. Boundas
Rosi Braidotti
Eugene Holland
Gregg Lambert
Dorothea Olkowski
Paul Patton
Daniel Smith
James Williams

Titles available in the series

Christian Kerslake, *Immanence and the Vertigo of Philosophy: From Kant to Deleuze*
Jean-Clet Martin, *Variations: The Philosophy of Gilles Deleuze*, translated by Constantin V. Boundas and Susan Dyrkton
Simone Bignall, *Postcolonial Agency: Critique and Constructivism*
Miguel de Beistegui, *Immanence – Deleuze and Philosophy*
Jean-Jacques Lecercle, *Badiou and Deleuze Read Literature*
Ronald Bogue, *Deleuzian Fabulation and the Scars of History*
Sean Bowden, *The Priority of Events: Deleuze's Logic of Sense*
Craig Lundy, *History and Becoming: Deleuze's Philosophy of Creativity*
Aidan Tynan, *Deleuze's Literary Clinic: Criticism and the Politics of Symptoms*
Thomas Nail, *Returning to Revolution: Deleuze, Guattari and Zapatismo*
François Zourabichvili, *Deleuze: A Philosophy of the Event* with *The Vocabulary of Deleuze* edited by Gregg Lambert and Daniel W. Smith, translated by Kieran Aarons
Frida Beckman, *Between Desire and Pleasure: A Deleuzian Theory of Sexuality*
Nadine Boljkovac, *Untimely Affects: Gilles Deleuze and an Ethics of Cinema*
Daniela Voss, *Conditions of Thought: Deleuze and Transcendental Ideas*
Daniel Barber, *Deleuze and the Naming of God: Post-Secularism and the Future of Immanence*
F. LeRon Shults, *Iconoclastic Theology: Gilles Deleuze and the Secretion of Atheism*
Janae Sholtz, *The Invention of a People: Heidegger and Deleuze on Art and the Political*
Marco Altamirano, *Time, Technology and Environment: An Essay on the Philosophy of Nature*
Sean McQueen, *Deleuze and Baudrillard: From Cyberpunk to Biopunk*

Forthcoming volumes

Ridvan Askin, *Differential Narratology*
Cheri Carr, *Deleuze's Kantian Ethos: Critique as a Way of Life*
Guillaume Collett, *The Psychoanalysis of Sense: Deleuze and the Lacanian School*
Ryan Johnson, *The Deleuze-Lucretius Encounter*
Marc Rölli, *Gilles Deleuze's Transcendental Empiricism: From Tradition to Difference* translated by Peter Hertz-Ohmes
Alex Tissandier, *Affirming Divergence: Deleuze's Reading of Leibniz*

Visit the Plateaus website at www.euppublishing.com/series/plat

TIME, TECHNOLOGY AND ENVIRONMENT
An Essay on the Philosophy of Nature

∽

Marco Altamirano

EDINBURGH
University Press

Edinburgh University Press is one of the leading university presses in the UK. We publish academic books and journals in our selected subject areas across the humanities and social sciences, combining cutting-edge scholarship with high editorial and production values to produce academic works of lasting importance. For more information visit our website: www.edinburghuniversitypress.com

© Marco Altamirano, 2016

Edinburgh University Press Ltd
The Tun – Holyrood Road, 12(2f) Jackson's Entry, Edinburgh EH8 8PJ

Typeset in Sabon by
Servis Filmsetting Ltd, Stockport, Cheshire

A CIP record for this book is available from the British Library

ISBN 978 0 7486 9157 9 (hardback)
ISBN 978 0 7486 9158 6 (webready PDF)
ISBN 978 0 7486 9159 3 (epub)

The right of Marco Altamirano to be identified as the author of this work has been asserted in accordance with the Copyright, Designs and Patents Act 1988, and the Copyright and Related Rights Regulations 2003 (SI No. 2498).

Contents

Acknowledgments	vii
Introduction: The Bifurcation of Nature	1

PART I: CRITIQUE OF THE BIFURCATION OF NATURE

1	The Clock and the Cogito	21
2	The Polarisation of Nature	48

PART II: TOWARD A NEW PHILOSOPHY OF NATURE

3	Difference and Representation: Deleuze and the Reversal of Platonism	83
4	Beyond the Nature-Artifice Divide: Technology, Milieu and Machine	117

Conclusion: The Technological Composition of Milieus	162
Index	171

For my daughter, Giselle

Acknowledgements

It's impossible to acknowledge all the people, events and creatures that contributed, however obliquely, intellectual ingredients to this book. Nonetheless, it's my privilege to be able to thank some friends and colleagues for their stimulating conversation, encouragement and support. I'd especially like to thank William McBride, John Sallis, Iain Hamilton Grant, Arkady Plotnitsky, John Protevi, William Zanardi, Patricia Curd, Nicolae Morar, Thomas Nail, Jim Bahoh, Deborah Goldgaber, Neal Miller, Michelle Landry, Somaieh Emamjomeh, Jason Wirth, Joe Hughes and Lee Nelson.

Fragments of this book were previously published in *Deleuze Studies and Foucault Studies*. I thank the editors of those journals for allowing me to reprint that material here. I'd also like to thank the Graduate School of Purdue University for supporting much of the initial research for this book through a Purdue Research Foundation Grant from 2011 to 2012.

I'm confident that many of the ideas presented here have been looted from the mind of my dissertation advisor, Daniel Smith, so I'd like to thank him in particular for his generosity. I'm also immensely grateful for Carol Macdonald at Edinburgh University Press, whose patience and expertise made the entire publication process such smooth sailing.

Introduction:
The Bifurcation of Nature

When Arthur Eddington discussed the problem of nature in his 1926 Gifford Lectures, he cast the problem in terms of that mainstay of philosophical furniture, the table.[1] For each table we see, he says, there is a second table that we do not see. The first table, of course, we are all familiar with – it is the hard, wooden, brown and typically unremarkable table. But there is another table that we can imagine given some pictures from physics. This other table is not brown, or any colour for that matter, and while it supposedly cohabits the same space as the first table, it does not have the same shape because it is mostly empty space and floating electrical charges. This second, physically imagined table is the bare table-object, stripped of any appearance or relation to subjects. And this physical table is located within a world full of physical entities like atoms, protons and electrons, whereas the familiar table inhabits a world full of buildings, books, humans and other creatures. Curiously, it is as if there were two distinct *natures* drawn along the lines of these two tables. First, there is a nature of flora and fauna, of tables and even of history, of political geographies, of wills and mankind. But there is also a physical-scientific nature, which may be conceived as more real, a nature of atoms and electrons, of basic physical elements and processes. And if this second, scientific nature is more real, then the first nature of creatures, trees, geographies and mankind seems to be a kind of illusion – we see tables, yet behind what we see lies a true world, arrived at not through our eyes but through our minds.

In this awkward caricature of physical nature, we are condemned to illusion by our very bodies – we may get through our days with the help of our senses, but in truth they deceive us at every turn. So while the senses provide us with useful cues for manoeuvring through the world, they deliver almost nothing when it comes to knowledge about the reality outside of our myopic subjectivity. But 'almost nothing' is not completely nothing: the senses do in fact serve an instrumental purpose when it comes to the production of scientific knowledge. After all, physical reality must be detected empirically, through bodies, charts, sensors and other technologies. But, fortunately, abstract physical theories are independent from the subjective conditions that are indispensable to their achievement. Physicists therefore display a characteristic aloofness with regard to philosophical questions about truth and knowledge. The truth, like a new species of frog, is simply out there to be discovered. When the physicist is in the laboratory, there is no need to consider the practices and institutions that form the context within which that laboratory exists, just as there is no need for the scientist to consider her mind and eyes, for the truth is independent of all such 'social' and 'subjective' concerns. Furthermore, this distinction between the subjective and the objective is definitive not only for physics, but for the natural sciences in general: the objectivity of scientific truth is inversely proportionate to the absence of any subjectivity that the inquirer might bring to it. Scientific results are thus rendered insensitive, apolitical, ahistorical, autonomous and human-independent. While this is surely an awkward caricature, it nonetheless informs the popular meaning of scientific 'objectivity'.

Science and Philosophy

While the physical sciences enjoy a privileged access to objective nature, it remains difficult to recognise ourselves in this scientific picture of nature. Perhaps that is why philosophy remains muddled in questioning the subject's place in that objectivity. As is well known, the distinction here between the objectivity of science and the subjectivity of philosophy has a long history. In fact, it enjoys a kind of preconfiguration in ancient Greek philosophy, when science and philosophy were not yet distinct enterprises. While the ancients didn't examine it in terms of subjectivity and objectivity, the problem of appearance and reality is a recognisable starting point for the distinction between objective knowledge and subjective opinion. So the problem, broadly conceived, dates back at least to Parmenides' two ways, truth and opinion (*aletheia* and *doxa*). Opinion, like appearance, changes, while truth stays the same eternally. While the many are feverishly swayed by opinion, the few, the philosophers, can walk along the steady path of truth. After Parmenides, Platonism will be interpreted as treating the problem of appearance and reality by splitting the world into two realms – there is a sensible realm of change and becoming, where illusion and semblance are dominant forces, and there is also an intelligible realm of self-same, eternal

Introduction

ideas, a realm of truth and unchanging essences (I hasten to add that, while this may be a common characterisation of the legacy of Platonism, Plato's dialogues themselves resist such facile two-world caricatures).

Time and movement are common elements throughout these dichotomous conceptions of truth. With the notable exception of Heraclitus, appearance has been understood as constantly shifting and never staying the same, while truth is steadfast and eternal. Heidegger interpreted Heraclitus' famous aphorism, 'Nature loves to hide', as stating that Being unveils itself as it veils itself.[2] Perhaps reality, in order to remain the same, must remain hidden behind a constantly changing appearance, making reality hidden amid its very appearance. But whether the problem of reality and appearance is understood in terms of truth and opinion or subjectivity and objectivity, it nonetheless remains a profoundly temporal problem – objective truth is still cast as eternal and unchanging. Thus whenever the question of truth arises, a question of time and movement is not far away. When Alfred North Whitehead famously wrote that 'the safest general characterisation of the European philosophical tradition is that it consists of a series of footnotes to Plato', he had in mind, at bottom, a distinction between the eternity of being and the dynamism of becoming (which conditions the subsequent notions of reality and appearance).[3] And one could make a strong argument in support of this characterisation of the history of Western philosophy by following the distinction between appearance and reality through its seventeenth-century modification into primary and secondary qualities (geometrical qualities of objects and sensible qualities, respectively), before reaching its Kantian transformation into things-in-themselves and things-for-us.[4] Of course, the distinction has undergone such immense contextual modifications throughout its history that it becomes difficult to maintain that it is even the same distinction throughout.[5] Nonetheless, the first part of this book provides some conceptual and historical support for the thesis that the divisions between an apparent, subjective, or changing world of becoming and a real, objective, eternal world of being hinge upon a similar conception of time and nature.

To be sure, the distinction between the natural sciences and philosophy does not correspond to the distinction between reality and appearance, or being and becoming. However, both science and philosophy have drawn the image of nature by means of a distinction between the sensible and the intelligible, ephemeral appearances and an eternal truth. In *Science and the Modern World*, Whitehead finds a European theological motivation behind modern science, since 'something more is wanted' for science 'than a general sense of the order in things'.[6] For Whitehead, science demands an

> inexpugnable belief that every detailed occurrence can be correlated with its antecedents in a perfectly definite manner, exemplifying general principles. Without this belief the incredible labours of scientists would be without hope. It is this instinctive conviction . . . which is the motive power of research – that there is a secret, a secret which can be unveiled . . . there seems but one source for its origin. It must come from the medieval insistence on the rationality of God, conceived as with the personal energy of Jehovah and with the rationality of a Greek philosopher.[7]

It was the rationality of the European deity that guaranteed, in the first instance, the intelligibility of nature, which thereby justified the rational search for the principles of nature. What is interesting about Whitehead's analysis here is not only the claim that European theology motivated modern science, but also that, as a matter of *conceptual procedure*, modern science begins with a belief in the intelligibility of nature. This implies that modern science approaches nature in terms of knowledge, in terms of intelligibility, which by definition is opposed to appearances and the senses. The contest between philosophy and science to lay claim to the intelligibility of nature is secondary to the sheer identification of nature with pure rationality over and against the obscurity of the senses. It is this procedural postulate that allows science to lay claim to nature – the nature we experience is a confused, multifarious and irrational nature of *doxa*, and it is to be judged against the intelligible, unified and rational standard of nature – the truth of nature cannot be experienced but only known, and science or *scientia* is precisely that knowledge of nature. Philosophy (before its separation from natural science) inaugurated the distinction between appearance and reality in order to lay claim to the eternal truths of nature. Ironically, the distinction ultimately facilitated the development of modern science, which subsequently laid claim to precisely the eternal truths sought by philosophy through the distinction. Yet, while the last century offers lamentably copious animosities and misunderstandings between them, both science and philosophy profoundly agree that appearance is at odds with reality.

Introduction

Two Tables: Materialism and Idealism

In his lectures, Eddington asks himself the question, 'you speak paradoxically of two worlds. Are they not really two aspects or two interpretations of one and the same world?'[8] Presumably, we should be able to identify the real table arrived at through scientific knowledge as the same wooden table that we sit in front of in everyday experience. But if the physical table is made up of empty space and floating particles, how does it become the solid brown table? How does the objectively real table become the subjectively apparent table? Eddington playfully muses that this must be the work of an 'alchemist Mind who transmutes the symbols' of physical process:

> The sparsely spread nuclei of electric force become a tangible solid; their restless agitations becomes the warmth of summer; the octave of aethereal vibrations becomes a gorgeous rainbow. Nor does the alchemy stop here. In the transmuted world new significances arise which are scarcely to be traced in the world of symbols; so that it becomes a world of beauty and purpose – and, alas, suffering and evil.[9]

This is surely a muddle. If the mind effects this artistic transmutation from an insensible (albeit detectable) physical table to the familiar table, then the mind would indeed be a strange and mysterious alchemist.[10] If, on the other hand, it is physical process itself that effects this transmutation, then an explanation seems conceivable if not forthcoming: the table must simply climb up the scientific ladder, through chemistry, biology, and even through cognitive, perceptual and psychological sciences in order to finally become the familiar table. Along its climbing path, we may perhaps require some supplementary fields to bridge the gap between physics and perception, but eventually we will be able to dispel the belief in the alchemist mind and finally understand how atoms and molecules become the table upon which we work, made in a certain year by a certain person, a period table of a certain architecture and wood. This would satisfactorily banish all subjective extravagancies, leaving behind only the pure, distilled, physical and real world. Alas, it remains difficult to recognise ourselves in this purely physical world, and history and experience stubbornly insist that there must be more to the story.

This is the basic dilemma we are left with: either the objective truth of the world is purely physical, in which case subjective experience is an illusion; or subjective experience is somehow real, in which case the objectivity of nature might be compromised by the admission of impure subjective elements into the real. It is the age-old and ongoing debate between materialism – or the view that only matter, and most commonly physical matter, is real – and idealism, as the view that reality is fundamentally mental, and so immaterial. One of the central aims of this book is to draw a line perpendicular to the opposing dead-ends of materialism and idealism.

When examining the debate between materialism and idealism, the first thing that becomes apparent is that both positions divide nature into subjective and objective poles – there is a table as it is for-us and a table as it is in-itself. After all, there would be no need to reconnect Eddington's tables if they had not been split up to begin with. Whitehead calls this arrangement the *bifurcation of nature*. Historically, the attempts at reconciling the bifurcated poles of nature have been largely unproductive, ranging from the strongly materialist stance claiming that experience is illusory and meaningless to the strongly idealist claim that the subjective conditions of knowledge prevent even science from apprehending the real in-itself, which lies forever beyond our grasp. For the most part, the strongly materialist stance changes nothing about our quotidian beliefs and practices and the strongly idealist critique is swatted away like some philosophical gadfly. On the disciplinary level, the so-called 'science wars' are over: the precedent for understanding nature has been confirmed by the massive successes of physics, chemistry and biology, henceforth aptly named 'natural sciences'. Scientists and philosophers are free to pursue their divergent courses, although they may enjoy some polite correspondence and amusing encounters. Yet it remains that, despite their divergences, both science and philosophy share a problem – and more, a crisis – of nature.

The Crisis of Nature

In *We Have Never Been Modern*, Bruno Latour critiqued the separation between the human and the non-human and, relatedly, the separation between science and politics. 'Even the most rationalist ethnographer', Latour tells us, 'is perfectly capable of bringing together in a single monograph the myths, ethnosciences, genealogies, political forms, techniques, religions, epics and rites of the people she is studying.'[11] It would be curious, then, for an ethnographer to study the 'Moderns', who are faced with an ecological crisis that implicates chemists, carbons, meteorologists, glaciers, industrialists and heads of state, but who nonetheless consider that science is

Introduction

independent from politics because it studies a human-independent objective nature. In view of this crisis, of course, it is impossible to conceive of nature as mere *décor*, a world separate from human interaction. However, there persists a tendency to consider the theoretical nature that science is concerned with as something human-independent. But there can be no easy separation of the theoretical and the practical here. Climate change may not offend the strongly materialist picture of nature – atoms and molecules may certainly continue their physical business – but it presents the nature of flora and fauna, of humans and history, with catastrophic possibilities. Recognising that humans have become a chief factor in geological change, the International Commission on Stratigraphy is considering the 'Anthropocene' as a potential geological epoch, at the same hierarchical level as the Pleistocene and the Holocene. What this means, partly, is that what normally concerns natural science exclusively, i.e. the 'natural world' or the object of the natural sciences, now forcefully presents a dynamic critique of human practices, i.e. the 'social world' or the object of the social sciences and humanities.[12] But more profoundly, the crisis demands that we critically examine the disciplinary framework that determines not only the natural sciences but also the humanities and social sciences – namely, the framework that operates through major dichotomies such as nature/culture, human/non-human, science/politics, etc. It is now impossible to define the earth independently of the human. In this regard, though this book is chiefly a critical examination of the concept of nature, it is nonetheless concerned with the challenge the ecological crisis presents in so far as it demands that we reconsider how we think about nature. We can no longer afford to subscribe to the idea that nature is irreconcilably bifurcated between two poles, that objective physical process is independent from subjective wills and affairs, and that nature is the autonomous décor upon which human history unfolds. And it is the earth that now argues that the humanities can no longer be separated from the natural sciences, and that the natural sciences can no longer be seen as apolitical, ahistorical and purely objective. The environmental crisis demands that we retrace the composition of nature beyond the subjective and objective poles.

And so we discover that the bifurcation of nature is not simply a conceptual affair, it is at once a political and a historical challenge. Instead of viewing themselves as detached commentators upon nature, philosophers and scientists must now reconsider their practices as active roles in the stewardship of the earth. Perhaps part of this challenge may be to reengage the philosophy of nature in a new way, beyond the bifurcated formulations of the past. Science has understood its success according to the postulate that it is, in fact, something independent, autonomous, objective, unified and abstract that we have been measuring with such success. Yet science would lose none of its success or importance (in fact, this would be better understood!) if we were to recognise it as fully integrated with human practices and the earth. Thus, part of the effort here is to argue that the picture of an independent, unified and objective nature is not innocuous – it is a political picture as much as it is a scientific one.

The Distribution of Nature

Some important moments in the history of the concept of nature will be examined in subsequent chapters, but for now we may note that the conceptual distribution of nature from the seventeenth century on has been cast in terms of (1) the subject (that 'alchemist mind') in its familiar world of familiar objects; (2) the objective physical process that supposedly subtends that world; and (3) the relation between them. At once we notice that this distribution is drawn along the lines of the modern subject, the subject of secondary appearances arising from primary qualities. Without the centrality of the modern subject, the distribution of natural qualities must be traced differently. Thus, the scholastics, for instance, did not distribute nature along the lines of a subject (as is the case with the primary-objective and secondary-subjective qualities distinction). Instead, as Robert Pasnau reports, they distributed nature alongside (1) basic or 'primary' qualities such as Hot, Cold, Wet and Dry; (2) non-basic tactile qualities such as heavy, light, hard, soft, rough and smooth; (3) other sensible qualities such as colour, sound, smell and taste; (4) occult qualities such as magnetism; (5) spiritual qualities such as light or colour as it exists in sensation; and (6) completely immaterial qualities such as thoughts and volitions.[13] It is unsurprising, then, that the distribution of nature traced upon the lines of the modern subject coincides with the modern revolution in science beginning in the seventeenth century. The modern subject's distribution of nature has been of massive service to science, as science operates by abstracting from how things appear to us in order to account for how objects relate to one another (there would not have been a revolution, of course, if Copernicus considered the sun only as it appeared to him, rising and falling). Unfortunately, this same distribution has caused philosophy considerable trouble. The relation between the subject and the object is a monstrous riddle, both epistemologically and metaphysically. Epistemologically, it is forbiddingly difficult to determine how the subject gains access to the object, if it does so at all; and metaphysically, it is exceedingly difficult to determine where the subject ends and where the object begins. The space between Eddington's two tables seems impossible to bridge. Yet we are at the dawn of an era where we cannot help but wonder how we ever considered subjects independently of objects, humans independently of nature, politics independent of the earth.

Still, Eddington's two tables remain all too familiar to us. Scientific knowledge popularly maintains its abstract and independent character, an objectivity that is supported by abstracting from the peculiarities of the inquirer and her environment. Whitehead claims the problem is that abstract scientific knowledge of nature is produced from the very concrete phenomena of nature, but it is recast as a substratum of nature, as if what is produced at the end of a long chain of concrete events actually came first, having been there all along. This is what he called the 'fallacy of misplaced concreteness', since it takes the abstract (scientific knowledge) and (mis)places it underneath the concreteness of real experience. Of course, Whitehead does not deny that there are concrete physical processes at work in nature, all he claims is that the abstract discursive pictures of these processes should be taken for what they are – abstractions from the concrete process of reality. But in order to better understand what Whitehead means by the fallacy of misplaced concreteness, we may profitably survey a few of the moments where he discusses it.

The Fallacy of Misplaced Concreteness

In *Process and Reality*, Whitehead writes:

> Philosophical thought has made for itself difficulties by dealing exclusively in very abstract notions, such as those of mere awareness, mere private sensation, mere emotion, mere purpose, mere appearance, mere causation. These are the ghosts of the old 'faculties', banished from psychology, but still haunting metaphysics. There can be no 'mere' togetherness of such abstractions. The result is that philosophical discussion is enmeshed in the fallacy of 'misplaced concreteness'.[14]

The fallacy, as presented in this passage of *Process and Reality*, seems to be a critique of abstractions that strip away metaphysical 'togetherness' or coherence by being overly divisive, in the sense that the terms are abstracted from their relations in order to preserve their 'mereness' or purity. In his earlier book, *The Concept of Nature*, Whitehead offers a different and considerably more complicated treatment of the 'long misconception of the metaphysical status of natural entities'[15] based on 'three components in our knowledge of nature, namely, fact, factors, and entities . . .':

> Fact is the undifferentiated terminus of sense-awareness; factors are termini of sense-awareness, differentiated as elements of fact; entities are factors in their function as the termini of thought . . . the immediate fact for awareness is the whole occurrence of nature. It is nature as an event present for sense-awareness, and essentially passing . . . Thus the ultimate fact for sense-awareness is an event.[16]

This passage should be parsed out slowly. First, Whitehead calls our attention to what we might call experience or sense-awareness, the basic sense of existing and perceiving, without differentiating exactly what we are perceiving. This basic and undifferentiated object of awareness is what Whitehead calls 'fact' – it is the simple and dumb sense of life happening. Of course, above and beyond the basic sense of awareness, we are aware of certain things – this book, this pizza, that electrode, that signal – and those elements of fact are what Whitehead calls 'factors'. Now, these factors, which are differentiated objects of awareness, can be abstracted from sense-awareness, and this function constitutes a different kind of awareness called 'thought'. Finally, when factors function in thought they become entities – like greenness, food, electricity and information – because they function separately from factors in sense-awareness (i.e. entities are abstractions from factors). There are no entities for sense-awareness precisely because entities are arrived at through a procedure of abstraction from sense-awareness resulting in thought. In reality, of course, fact, factors and entities are all part of the whole process of nature. But Whitehead's ultimate point is that nature is essentially a process *in occurrence* – it is not primarily entities or even stuff that fills the universe like sand fills a box. Thus, the ultimate fact for sense-awareness is an event, since the object of our awareness, before it is differentiated into factors that are subsequently abstracted into entities, is simply an event occurring.

This point cannot be stressed enough. We are so accustomed to

Introduction

thinking of nature as a spatial entity, a kind of *res extensa* that is populated with various substances, that it is only with considerable difficulty that we can focus on its basically temporal character. But once we discover nature as a primarily temporal process, we find that it is the passage of time itself (and not primarily space) that is consciously modified in order to produce either the general scientific table or the particular brown table that Eddington contemplates. The point here is that the bifurcation of nature is a procedure of thinking rather than a substantial bifurcation – nature is essentially a process and not a substance that can be split into subjective and objective poles. Thus the fallacy of misplaced concreteness is a fallacy of bifurcating nature, since nature is split into objective physical reality and subjective appearance only by abstracting entities from the process of reality. Whitehead elaborates this point when describing how the fallacy of misplaced concreteness is committed:

> The entity has been separated from the factor which is the terminus of sense-awareness. It has become the substratum for that factor, and the factor has been degraded into an attribute of the entity. In this way a distinction has been imported into nature which is in truth no distinction at all. A natural entity is merely a factor of fact, considered in itself. Its disconnexion from the complex of fact is a mere abstraction. It is not the substratum of the factor, but the very factor itself as bared in thought. Thus what is a mere procedure of mind in the translation of sense-awareness into discursive knowledge has been transmuted into a fundamental character of nature. In this way matter has emerged as being the metaphysical substratum of its properties, and the course of nature is interpreted as the history of matter.[17]

In other words, the abstraction by which we consider physical process is a procedure of the mind that disconnects entities from the passage of nature, which is the fundamental fact of nature. Unfortunately, an illusion arises because once a factor of nature is abstracted in thought it may subsequently be recast as the substratum of the factor. Thus, the abstract, which is a result of the translation of sense-awareness, is misplaced underneath the concrete as its substratum. The critical point to note here is that the fallacy of misplaced concreteness is directed toward discursive knowledge that reconfigures nature in terms of space and matter, overlooking the foundation of that very knowledge in the passage of nature. At the basis of physical materialism, then, there is a disregard for the fundamental passage of nature.

In this regard, Whitehead's admiration for Bergson, who also argued against the spatial interpretation of concreteness, comes as no surprise. In a passage of *Matter and Memory* that (with some stylistic changes) could pass for a passage out of Whitehead's *Process and Reality*, Bergson offers his own diagnosis of the fallacy of misplaced concreteness:

> Amorphous space, atoms jostling against each other, are only our tactile perceptions made objective, set apart from all our other perceptions on account of the special importance which we attribute to them, and made into independent realities – thus contrasting with the other sensations which are then supposed to be only the symbols of these. Indeed, in the course of this operation, we have emptied these tactile sensations of a part of their content; after having reduced all other senses to being mere appendages of the sense of touch, touch itself we mutilate, leaving out everything in it that is not a mere abstract or diagrammatic design of tactile perception: with this design we then go on to construct the external world. Can we wonder that between this abstraction, on the one hand, and sensations, on the other hand, no possible link is to be found? But the truth is that space is no more without us than within us . . . Concrete extensity, that is to say, the diversity of sensible qualities, is not within space; rather is it space that we thrust into extensity.[18]

Here, Bergson brilliantly highlights the fallacy of misplaced concreteness at the root of the bifurcation of nature. If we return to the example of Eddington's two tables, the problem with the bifurcation of nature is that the familiar brown table and the table made of empty space and floating particles are supposedly both tables in front of Eddington. In other words, the objectively fathomed physical table has been inserted underneath the familiar table as a substratum, when in fact it is a 'diagrammatic abstraction' from the diversity of sensible qualities, which Bergson calls 'concrete extensity'. Concreteness, then, is the concreteness of experience in time – which helps to explain the provocative ending to the above passage in Bergson, 'concrete extensity . . . is not within space; rather is it space that we thrust into extensity'. In other words, space (as the imagined container or *res extensa* subtending the objects we encounter through sense-awareness) is an abstraction that we reinsert into concrete extensity (a point we will return to in Chapter 2 below).

Bergson and Whitehead are in such precise agreement when criticising the metaphysics of misplaced concreteness because they both agree that time should be prioritised in the philosophy of nature. When Whitehead was developing his account of nature and matter in the early Tarner Lectures in 1919, he acknowledged his affinity with Bergson's temporally based analysis of matter: 'I believe

Introduction

that in this doctrine I am in full accord with Bergson, though he uses "time" for the fundamental fact which I call the "passage of nature".'[19] Whitehead also announced his agreement with Bergson's critique of the spatialisation of matter in his Lowell Lectures of 1925 that became the book, *Science and the Modern World*, although by then he also admits a distinction between Bergson's criticism and his own:

> This simple location of instantaneous material configurations is what Bergson has protested against, so far as it concerns time and so far as it is taken to be the fundamental fact of concrete nature. He calls it a distortion of nature due to the intellectual 'spatialisation' of things. I agree with Bergson in his protest... there is an error; but it is merely the accidental error of mistaking the abstract for the concrete. It is an example of what I will call the 'Fallacy of Misplaced Concreteness.'[20]

Given the ubiquity of the error, Whitehead seems overly generous here when claiming that it is merely an 'accidental error'. For his own part, as Whitehead recognises, Bergson found the error to be systematic – when the intellect considers perception, it takes it out of its inherently eventual (or temporal) dimension and relocates it within a spatial framework. The intellect thereby considers human perception to be something like a camera, providing a 'photographic view of things, taken from a fixed point by that special apparatus which is called an organ of perception'.[21] Our ocular apparatus is henceforth depicted as receiving material disturbances from a location in space at a given instant that subsequently pass through the nerves and arrive at the brain. At this point in the story, Bergson discovers a fantastic transmutation resulting in a purely mental representation of the surrounding world:

> Here I am confronted by a transformation scene from fairyland. The material world which surrounds the body, the body which shelters the brain, the brain in which we distinguish centres, [the philosopher] abruptly dismisses, and, as by a magician's wand, he conjures up, as a thing entirely new the representation of what he began by postulating. This representation he drives out of space, so that it may have nothing in common with the matter from which he started. As for matter itself, he would fain go without it, but cannot, because its phenomena present relatively to each other an order so strict and so indifferent as to the point of origin chosen, that this regularity and this indifference really constitute an independent existence.[22]

Here the world is mirrored into a double by perception – the world 'out there' is taken as a spatial box of objectivity while the mind 'inside' becomes a subjective illusionist, an illustrator of representations. Consequently, the qualities that conscious perception 'adds on' to the basic physical process must be explained. But for the scientific materialist, to explain these conscious qualities would require their reconnection to the bare physical process that one arrived at *only by abstracting from those very qualities*. For the idealist, on the other hand, we cannot begin with mental representations and attempt to deduce a purely material order of things that was excluded from the outset. Small wonder, then, that there is a dilemma.

There is no way of escaping the dilemma of reconnecting the mind to the world in so far as we persist in the metaphysics of misplaced concreteness and the bifurcation of nature that it generates. Nonetheless, by examining the root of the dilemma, we have discovered a possible way out of it. The bifurcation of nature is a thoroughly spatial picture of nature – a subject confronts a perceptual object in space like Eddington confronts his familiar table, and considers it the mere appearance of some deeper physical reality, yet he preserves the perceptual space and imports it into the picture of this deeper physical reality, and can thereby maintain that there are in fact two tables, one which is familiar, subjective and illusory, and one which is scientific, objective and real. An escape route from this dilemma would require a reconfiguration of the concept of nature that avoids the pitfalls of Whitehead's fallacy and what Bergson calls 'intellectual spatialisation'. And if Bergson and Whitehead are correct, the most promising approach toward the philosophy of nature is to think of time (and not perceptual space) as the fundamental datum of nature.

The central thesis of this book is that the concept of nature is more productively formulated in terms of time and events, rather than in terms of subjects and objects in space, as the modern bifurcation of nature holds. The first part of the book (Chapters 1 and 2) forms a kind of introduction to the problem of nature and how it has been formulated historically and conceptually, while the second part (Chapters 3 and 4) offers conceptual resources (such as difference, becoming, technology, milieu and machine) for developing a new philosophy of nature based on time and events. Chapter 1 examines a pivotal moment in the history of the philosophy of nature, with the emergence of the modern subject during the seventeenth century and the beginnings of the separation of science from philosophy, physics from metaphysics. Chapter 2 considers Bergson's metaphysics of time and matter, which provides an alternative to the bifurcation of

Introduction

nature. In Bergson, mind and matter are not spatially bifurcated but rather polarised upon a continuum of time. By recasting the distinction between the subject and the object in terms of time, Bergson's polarisation of nature is an effort toward the reintegration of the mind and the body (as well as knowledge and experience). Chapter 3 begins a positive reformulation of the concept of nature by examining Deleuze's concept of difference in-itself alongside the concepts of simulacrum and becoming. While nature is typically understood as populated by subjects and objects, the concept of difference allows us to consider how identities such as subjects and objects are produced, not by other identities but rather by difference and becoming. Finally, Chapter 4 outlines how the concepts of technology, milieu and machine disrupt the disciplinary apparatus that produces knowledge of nature by opposing nature to artifice and technology. The temporally valorised concepts examined in the second part of this book ultimately draw the profile of a philosophy of nature that is beyond the nature-artifice divide precisely because it is based on time and events. A conclusion summarily discusses how regimes of knowledge are traced upon forms of technology, and the productive composition of the human through technology.

Notes

1. Arthur Eddington, *The Nature of the Physical World* (Whitefish: Kessinger Publishing LLC, 2010). In *A Thousand Plateaus*, Deleuze and Guattari mention Eddington's two tables in the context of what they call 'formal distinctions'. Gilles Deleuze and Félix Guattari, *A Thousand Plateaus: Capitalism and Schizophrenia*, trans. Brian Massumi (Minneapolis: University of Minnesota Press, 1987), p. 58.
2. For a magisterial analysis of the history of the idea of nature alongside Heraclitus' saying, 'phusis kruptesthai philei', see Pierre Hadot, *The Veil of Isis: An Essay on the History of the Idea of Nature* (Cambridge, MA: Harvard University Press, 2006).
3. A. N. Whitehead, *Process and Reality (Corrected Edition)*, ed. David Ray Griffin and Donald Sherburne (New York: Free Press, 1978), p. 39. He continues, 'the things which are temporal arise by their participation in the things which are eternal. The two sets are mediated by a thing which combines the actuality of what is temporal with the timelessness of what is potential' (p. 40).
4. Perhaps the most immense modification of the distinction arrives with Kant, who, after being awoken from his dogmatic slumber by Hume's devastating critique of causality, configures the in-itself as

beyond the conditions of knowledge 'in order to make room for faith'. Immanuel Kant, *Critique of Pure Reason*, trans. and ed. Paul Guyer and Allen Wood (Cambridge: Cambridge University Press, 1998), p. 117 (Bxxx).
5. In this regard see Mi-Kyoung Lee, 'The Distinction between Primary and Secondary Qualities in Ancient Greek Philosophy', and Robert Pasunau, 'Scholastic Qualities, Primary and Secondary', both in Lawrence Nolan (ed.), *Primary and Secondary Qualities: The Historical and Ongoing Debate* (Oxford: Oxford University Press, 2011).
6. A. N. Whitehead, *Science and the Modern World* (New York: Free Press, 1967), pp. 12–13. In an unfortunate Eurocentric display, Whitehead muses that 'Chinese science is practically negligible' because the Chinese did not enjoy a belief in a rational deity, which is why 'there is no reason to believe that China if left to itself would have ever produced any progress in science. The same may be said of India.' *Science and the Modern World*, p. 6.
7. Whitehead, *Science and the Modern World*, p. 12.
8. Eddington, *The Nature of the Physical World*, p. xii.
9. Eddington, *The Nature of the Physical World*, p. xv.
10. John Protevi relates Susan Hurley's playful casting of this arrangement as 'the classical sandwich model of the mind'. See Susan Hurley, *Consciousness in Action* (Cambridge, MA: Harvard University Press, 1998), p. 401. Protevi elaborates Hurley's culinary metaphor of the 'individual as rational cognitive subject' with the following depiction: '[the subject] gathers sensory information in order to learn about the features of the world; processes that information into representations of those features; calculates the best course of action in the world given the relation of those represented features of the world and the desires it has . . . and then commands its body and related instruments to best realise those desires given the features of the world it has represented to itself.' John Protevi, *Political Affect: Connecting the Social and the Somatic* (Minneapolis: University of Minnesota Press, 2009), p. 3.
11. Bruno Latour, *We Have Never Been Modern*, trans. Catherine Porter (Cambridge, MA: Harvard University Press, 1993), p. 7.
12. On this point, see Bruno Latour's 2012–2013 Gifford Lectures, 'Facing Gaia: A New Enquiry into Natural Religion', available at <http://www.ed.ac.uk/schools-departments/humanities-soc-sci/news-events/lectures/gifford-lectures/archive/series-2012-2013/bruno-latour> (last accessed 20 August 2015).
13. Pasnau, 'Scholastic Qualities, Primary and Secondary', p. 43.
14. Whitehead, *Process and Reality*, p. 18.
15. A. N. Whitehead, *The Concept of Nature* (Charleston: BiblioBazaar Reprints, 2007), p. 25.
16. Whitehead, *The Concept of Nature*, p. 24.

Introduction

17. Whitehead, *The Concept of Nature*, p. 25. Although this passage is taken from *The Concept of Nature*, Whitehead's Tarner Lectures of 1919, the same fallacy is treated in *Process and Reality*, which is a transcription of Whitehead's Gifford Lectures of 1927–1928, which followed upon the heels of Eddington's own 1926–1927 Gifford Lectures.
18. Henri Bergson, *Matter and Memory*, trans. N. M. Paul and W. S. Palmer (New York: Zone Books, 1991), pp. 216–17.
19. Whitehead, *The Concept of Nature*, pp. 52–3.
20. Whitehead, *Science and the Modern World*, pp. 50–1. Whitehead disagrees that this mistake is 'a vice necessary to the intellectual apprehension of nature'.
21. Bergson, *Matter and Memory*, p. 39.
22. Bergson, *Matter and Memory*, p. 39.

PART I
Critique of the Bifurcation of Nature

1
The Clock and the Cogito

The bifurcation of nature (henceforth referred to as simply 'bifurcation') refers to a view that splits the world into a familiar world of flora and fauna and a scientific world of floating particles and empty space. As described in the introduction, bifurcation subtends the long-standing debate between materialism and idealism, with its correlate dilemma regarding the connection between mind and matter. But why was nature bifurcated in the first place? What was at stake? In order to appreciate the reasons for separating the mind from matter, we must review a pivotal moment in the history of the philosophy of nature. In this chapter, we examine, first, the epistemological commitments subtending the bifurcation of nature into primary and secondary qualities in Descartes and seventeenth-century natural philosophy, and, second, the concept of time that subtends the bifurcation and the mechanistic picture of the world.

The Problem of Error

Descartes famously begins his *Meditations* with an intellectual autobiography of sorts. He confesses that many of his opinions had been discovered to be false and, with the passage of years, he had allowed numerous falsehoods to accumulate. Noticing an opportune time, he decides to free his mind of cares, withdraw into solitude, and raze all of his past opinions 'in order to establish something stable and lasting'.[1] His celebrated goal is to ground knowledge upon a firm and absolute foundation, something indubitable, something certain.

Gilles Deleuze once said that the problem for seventeenth-century philosophy was the problem of error.[2] In Descartes, this problem is an obsession – everywhere we see him in pursuit of certainty, developing principles and rules to protect himself from falsehood (the *Discourse on Method*, Descartes' first publication, was originally titled, 'Discourse on the Method of rightly conducting one's reason and seeking the truth in the sciences, and in addition the Optics, the Meteorology, and the Geometry, which are essays in this Method').

As we shall see, the problem of error, with its attendant configuration of knowledge and search for certainty, profoundly determines Descartes' philosophy of nature.

Descartes finds that the best way of securing certainty is to vigilantly prevent falsehood from entering into his thinking. As a stay against falsehood, he deploys his well-known method of doubt. It is a curious method, though, because it starts backwards, as it were: in order to decide whether or not something is true, you first must pretend it is false. So, for example, in order to decide whether or not I really exist, I make the claim, 'I do not exist'. Suddenly, I realise that I've involved myself in a contradiction since, in order to make the claim, I must affirm the 'I' that I also deny. In other words, I must posit the existence of myself, 'I', in order to deny that 'I' exist. In this fashion, the principle of non-contradiction becomes the logical guarantor of the method of doubt: what is beyond all doubt is what cannot be doubted without admitting a contradiction. One can be *absolutely* certain of something if and only if doubting it entails a contradiction – that is the criterion of absolute certainty. Hence, Descartes starts out by negating everything – denying the existence of the cogito, God and the world – only in order to reaffirm them.

While this method might initially seem rather innocuous (perhaps a little curious, but not unreasonable), it actually changes the way we question and understand nature. Before Descartes, the focal point of scholarly life was the book, the explication of venerable texts. The way to produce knowledge, including knowledge about nature, was to examine ancient and dusty books. In order to discover, for example, why the sun rises, why the stars shine, why it rains, or why fire moves up and rocks fall down, one would look at an ancient book, perhaps Aristotle's *De Caelo*, and one would learn that rocks fall down because the earth is their natural place (I, 8, 277b; I, 2, 269b). It would have seemed absurd, then, to produce knowledge about rocks by measuring their speed while falling, and even more absurd to repeatedly measure such a phenomenon, especially in public, for everyone to see and understand. But things were changing in the world – after Descartes the goal of scholarship is no longer to examine ancient books but rather to examine the book of the world, to make concrete observations and perform methodological experiments that everybody could understand.[3] Nature would no longer be something studied through books and lofty principles but rather something interrogated through instruments and phenomena. This new technique for producing knowledge of nature not only

The Faculty of Judgement

Descartes finds the cogito to be a clear and distinct idea because, again, it cannot be doubted without admitting a contradiction. But what kind of thing is the cogito? In order to answer this question, Descartes embarks upon the first of two remarkable exercises in imagination. He imagines a figure replete with hands, arms and a face, but curiously, he draws such a figure only to subsequently strip it of all its corporeality.[4] After all, I can doubt that I have hands, arms, legs or a face – I could very well be a brain in a vat, or simply an immaterial dream. There is nothing contradictory about doubting those corporeal parts, and in the absence of some divine guarantee, we cannot be certain that material bodies even exist. The existence of the cogito is certain without qualification from the body. Therefore, the body is not essential to the cogito. Thus, because of the premium placed upon certain (or indubitable) knowledge, the indubitable cogito is rendered substantially different from the dubious body. And along the way, the question, 'what kind of thing is the cogito' becomes 'what knowledge about the cogito is beyond all doubt?'

To be sure, the method of doubt is not purely negative, and there are positive elements of the cogito discovered in this exercise. The cogito is found to be 'a thing that thinks. What is that? A thing that doubts, understands, affirms, denies, is willing, is unwilling, and also imagines and has sensory perceptions.'[5] The cogito is therefore, positively, a substance endowed with a set of faculties. However, this positive definition of the cogito is traced through the omission of corporeal substance, the body and its worldly environment. That is to say, the affirmation of the essential elements of the cogito is arrived at only through the negation of the dubious, and therefore inessential, element of corporeality. Just as Descartes withdrew into solitude by freeing his mind of all worldly cares in order to raze his former opinions, he distils the cogito from its corporeal environment by eviscerating its connection to belief and appearance. The cogito is thus cast in an epistemic figure, in the sense that the epistemological criterion of certainty serves to define the cogito.

We can already see a natural division here between corporeal bodies and the incorporeal cogito, a natural division facilitated by an epistemological standard of certainty. The cogito is separated from extended substance as a clear and distinct idea is separated from its confused and obscure aspects. The body and the things and images that appeared within the world simply cannot satisfy the criterion of certainty, they are like so many opinions. The entirety of the world could be a dream, and Descartes 'can make judgments only about things that are known'.[6]

Unfortunately, after Descartes' epistemological culling of opinions, there is nothing left that is known outside of the cogito, so judgement does not have much to consider. Yet his mind continues to wander imaginatively, 'and will not yet submit to being restrained within the bounds of truth', which is to say, within the single clear and distinct idea of the cogito. It almost seems as if he is forced to allow his imaginative musing: 'very well then; just this once let us give it a completely free rein, so that after a while, when it is time to tighten the reins, it may more readily submit to being curbed'.[7] And in this way, he embarks upon his second remarkable exercise in imagination. Just as he arrived at the cogito by fleshing out an image of himself as an embodied man, only to strip away the imagined corporeality from the idea of himself, he now fills the world with new images in order to strip them of their confused and obscure parts so that he may discern what, if anything, is clear and distinct about material bodies. The body that he initially focuses his attention upon is the famous piece of wax.

Descartes strips the wax down to something hardly recognisable, something 'merely extended, flexible and changing', but realises that these abstractions cannot be the product of the imagination. After all, the extended and flexible wax can be changed into countless shapes, and we can understand that without imagining all of them.[8] There is something about this abstract wax that exhausts the imagination – so the imagination could not have identified this abstract entity. But what, then, gives us the wax? It is at this point in the narrative that we find Descartes looking out of his window, perhaps still in his day-robe, and it dawns on him that while he sees 'hats and coats which could conceal automatons', he *judges* that they are men.[9] It is judgement that discerns things, ranging from the wax to the cogito, and not the imagination or the senses. The wax could change in a variety of ways – it could assume the shape of a penguin, a shoe, or a graham cracker, for example. If considered by the imagination or the senses alone, these different configurations of the piece of wax would appear to be distinct things, yet judgement affirms it to be the same wax throughout its changes. Similarly, in the Sixth Meditation, Descartes reengages his insubordinate imagination in order to distinguish it from pure understanding by considering a triangle and a chiliagon, a thousand-sided figure: 'if I want to think of a chiliagon, although I understand that it is a figure consisting of a thousand sides just as well as I understand the triangle to be a three-sided figure, I do not in the same way imagine the thousand sides or see them as if they were present before me'.[10] When we imagine a triangle, we both *understand* it as having three sides as well as *imagine* the three sides. However, with a chiliagon we do not imagine the thousand sides, because imagination, and this is the essential point, *cannot count clearly*.[11] Try as you might, if you imagine a wall of ruddy bricks, particularly if you imagine a wall made up of a great many bricks, you cannot count the bricks in your imagination (and neither would you trust your eyes to count them). The understanding, however, can easily deal with large numbers. It is almost as if imagination is too close to the moving, corporeal realm, such that it is forced to accede some of that movement into its figures, rendering them unstable. Consequently, imagination only provides confused representations. The imagination, in other words, is a kind of halfway house between the confusion of changing sensibility and the discerning stability of understanding and judgement.[12]

The faculty of judgement affirms and denies, and it is responsible for affirming the identity of a substance that persists through time. So the cogito, for example, is identified through the affirmation of a set of mental faculties and the denial of the corporeal environment that surrounds it. In this sense, the cogito is separated from the world by judgement. Yet the cogito maintains several ideas about the world – that there is some wax burning in the room, that there are men outside the window. Do these ideas correspond to the world? Judgement is responsible for answering such questions. So after separating the cogito from the world, judgement incurs the burden of arbitrating the cogito's relation to the world – deciding whether or not the ideas within the cogito relate to things outside of the cogito, things in the world.

Descartes finds that 'the chief and most common mistake which is to be found here consists in my *judging* that the ideas which are in me resemble, or conform to, things located outside of me'.[13] Here it becomes clear that Descartes' epistemological problem of certainty (the problem of avoiding falsehood) is resolved through a bifurcation of nature into mental substances (like the cogito) and corporeal substances (like the world). We can be certain when judging clear and distinct ideas within the cogito, but we hazard error when judgement is illegitimately deployed in the world ('I can make judgements only about things that are known'). Thus, Descartes bifurcates nature in order to accomplish two related goals: (1) identify the source of error and (2) secure a method of certainty. The bifurcation of nature provides a jurisdiction for judgement, a legitimate province for judgement secured from the confusion and obscurity of the illegitimate territory of the sensible world. Descartes states this point more carefully in his *Principles* (I, 68) when distinguishing what can be clearly known regarding bodies and sensation:

> In order to distinguish what is clear in this connection from what is obscure we must be very careful to note that pain and colour and so on are clearly and distinctly perceived when they are regarded merely as sensations or thoughts. But when they are judged to be real things existing outside our mind, there is plainly no way of understanding what sort of things they are. If someone says he sees colour in a body or feels pain in a limb, this amounts to saying that he sees or feels something there of which he is wholly ignorant or, in other words, that he does not know what he is seeing or feeling. Admittedly, if he fails to pay sufficient attention, he may easily convince himself that he has some knowledge of what he sees or feels, because he may suppose that it is something similar to the sensation of colour of pain which he experiences within himself. But if he examines the nature of *what is represented* (repraesentet) *by the sensation* of colour or pain – what is represented as existing in the coloured body or the painful part – he will realise that he is wholly ignorant of it.[14]

A sensation may be clear and distinct simply as a sensation – thus far it is merely as a mode of thought, without referring to anything else. Descartes can safely claim that he is feeling a certain way or entertaining a certain thought but, if he tries to attribute it to some body outside of the cogito, then he is prone to err. The separation of the cogito from the world by the faculty of judgement thus incurs the problem of representation: we may very well see a wall of ruddy bricks and feel pain when it falls upon us, but we cannot say what is represented by the ruddiness or the pain without falling into error. For Descartes, sensible ideas are peculiar to the mind, and do not represent anything outside of the mind: 'sensations of tastes, smells, sounds, heat, cold, light, colours and so on – [are] sensations which do not represent anything located outside our thought'.[15] The entire sensible world is thereby locked within our minds.[16]

The Bifurcation of Nature into Primary and Secondary Qualities

As previously mentioned, Whitehead suggests that the modern scientific view of nature that emerged in the seventeenth century was encouraged by the belief in a rational deity that would guarantee an orderliness and intelligibility to nature. In Descartes we find support for this view: through the veracity of God we are able to secure some knowledge of the natural world outside of the cogito.[17] But first, nature must be reconfigured in order to conform to the standard of the knowing subject. In the following section, we examine how the bifurcation of nature into primary and secondary qualities is ultimately drawn along the contours of the knowing subject.

PRIMARY AND SECONDARY QUALITIES

Like Aristotle, Descartes was a plenist. There are no voids or empty spaces in corporeal substance, and the Cartesian world is a universal plenum (it is *res extensa* – extended thing). But unlike Aristotle, Descartes was a mechanist. Corporeal substance is made of shifting material bodies that interact through the mechanistic principle of action by contact, like a pedal turns a crank or a gear turns a clock dial. Descartes' plenism identifies corporeal substance with material bodies (there is no place or space without material bodies – extended space is material), and his mechanism entails that this material body is essentially the same – all material is the same throughout and is ruled by the same laws. The reason different bodies interact in diverse ways is because they have diverse geometrical qualities, so fire burns wood simply because of the motion of its parts – fire is made of sharp and violently moving little particles that separate even the tiniest parts of wood thus transforming it into fire, smoke and ash.[18] These geometric qualities – such as quantity, length, breadth, motion, depth, etc., in short, quantifiable qualities of bodies – are called primary qualities. Primary qualities thus determine and explain how bodies in motion relate to each other, independently of subjects. Qualities such as colour, smell, taste and sound – in short, sensible qualities – are called secondary qualities. They are purely mental, and express how material bodies relate to conscious subjects.

Already we find that nature is divided by reference to the subject: primary qualities are how bodies in motion relate to each other (independently of us), whereas secondary qualities express precisely the relation of bodies to us (but not to each other). In the *Principles*, Descartes outlines the process of sensation in order to argue that while primary qualities have dispositions that set up sensations, there is no way to comprehend how size, shape and motion would produce colour, and thus secondary qualities cannot be attributed to real bodies:

> Finally, let us consider heat and other qualities perceived by the senses, in so far as those qualities are in objects, as well as the forms of purely material things, for example the form of fire: we often see these arising from the local motion of certain bodies and producing in turn other local motions in other bodies. Now we understand very well how the different size, shape and motion of the particles of one body can produce various local motions in another body. *But there is no way of understanding how these same attributes (size, shape and motion) can produce something else whose nature is quite different from their own* – like the substantial forms and real qualities which many <philosophers> suppose to inhere in things; and we cannot understand how these qualities or forms could have the power subsequently to produce local motions in other bodies. Not only is all this unintelligible, but we know that the nature of our soul is such that different local motions are quite sufficient to produce all the sensations in the soul. What is more, we actually experience the various sensations as they are produced in the soul, and we do not find that anything reaches the brain from the external sense organs except for motions of this kind. In view of all this we have every reason to conclude that the properties in external objects to which we apply the terms light, colour, smell, taste, sound, heat and cold – as well as the other tactile qualities and even what are called 'substantial forms' – are, so far as we can see, simply various dispositions in those objects which make them able to set up various kinds of motions in our nerves <which are required to produce all the various sensations in our soul>.[19]

Sensations are caused by a relation between nerves (corporeal) and the soul (mental), so while the primary qualities of material bodies may be disposed to move the nerves in certain ways, there is no way to comprehend a supposed correspondence between a colour, for example, and its imputed object. This is due, in part, to a strictly mechanistic account of nature: sensations are products of nerves and the soul, so they do not really apply to the primary qualities that are moving the nerves. But it is also an epistemological account of nature: sensations cannot be known in the same way that primary qualities can be known *thus* they do not have the same nature. Primary qualities are purely geometric and intelligible in terms of mechanist principles. Secondary qualities are not geometric and mechanistic – they are not physically real, i.e. they exist purely in thought. The bifurcation of nature into primary and secondary qualities is thus drawn on the outlines of the knowing subject – secondary qualities lie within the subject and can be known only as modes of thought, while primary qualities lie outside of the subject and can be known as modes of extension or corporeal substance.

THE PINEAL GLAND

But if secondary qualities do not represent objects outside of the mind, if they are merely qualities that reside in our minds, then how do primary qualities cause, promote or dispose (however it may be) secondary qualities? If there are no red or ruddy bodies in nature, then how is it that when I come across material bodies that have the primary qualities of bricks, I perceive ruddiness? What strange alchemist is responsible for forging this connection between primary and secondary qualities in order to bring about conscious experience? This question becomes an immense objection to Descartes because it is not at all clear how, for example, when I drink a martini (which is purely material) I experience some clearly mental effects, especially considering that the soul is supposed to be immaterial (that is, independent of material). Thus, the question about how secondary qualities relate to primary qualities becomes a question of how the soul, which is non-extended, relates to the extended body.

The union of the soul and the body is the chief topic of the celebrated correspondence between Descartes and Princess Elisabeth of Bohemia.[20] In the *Passions of the Soul*, dedicated to Elisabeth as the culmination of this correspondence, Descartes approached the question of the relation between the soul and the body in terms of extension:

> And the soul is of such a nature that it has no relation to extension, or to the dimensions or other properties of the matter of which the body is composed: it is related solely to the whole assemblage of the body's organs. This is obvious from our inability to conceive of a half or a third of a soul, or of the extension which a soul occupies. Nor does the soul become any smaller if we cut off some part of the body, but it becomes completely separate from the body when we break up the assemblage of the body's organs.[21]

Here Descartes states that the soul cannot be understood in terms of extension because of its essential indivisibility – which simply reiterates the dualism between mental substance and corporeal substance. However, given the clear perception of sensible secondary qualities arriving from parts throughout the body, there is manifestly some relation between the soul and the body. But Descartes' substantial dualism is strictly upheld – despite their union, the body maintains its corporeal nature (it can be extensively split) and the soul maintains its non-extended nature (it remains intact when a body is broken up). So it might seem surprising, then, that Descartes proceeds to pinpoint *in extension* the site of the relation between the non-extended soul and the extended body:

> Let us therefore take it that the soul has its principal seat in the small gland located in the middle of the brain. From there it radiates through the rest of the body by means of the animal spirits, the nerves, and even the blood, which can take the impressions of the spirits and carry them through the arteries to all the limbs.[22]

Perhaps even more surprisingly, some of Descartes' contemporaries criticised the idea that the soul was located in the pineal gland, but not always on the grounds that it located something immaterial and non-extended within something material and extended. The grounds for dispute often seemed to lie elsewhere, most notably with critics disputing the particular location of the soul's inherence within the body. The physician Thomas Willis, for example, criticised Descartes' soul-imbued pineal gland because 'animals, which seem to be almost quite destitute of Imagination, Memory, and other superior Powers of the Soul', have this organ 'large and fair enough'.[23]

While it seems difficult to understand Descartes' identification of the pineal gland as the seat of the soul, we can nonetheless understand the necessarily hypothetical approach he took toward the question of the union between the mind and the body. If the union were conceivable as a mode of extension, then the soul, like the body, would be a corporeal substance. But if the union were conceivable as a mode of thought, then the body, like the soul, would be a mental substance. There is manifestly a relation, it is felt and may even be located, but explaining this relation would be tantamount to explaining the relation between primary and secondary qualities, and that explanation, for the reasons above, is impossible. Knowledge of mechanism is the purchase of redrawing the concept of nature along the contours of the knowing subject, but it comes at the cost of irrevocably separating mind from matter.

The Clock and the Cogito

For the ancient Greeks, *mēkhanē* denoted ways of tricking nature – in particular, ways to make things move by artificial means in ways that seem contrary to their nature. Naturally, for instance, a heavy object is heavy to lift, but with the use of levers, nature can be tricked and the heavy can become light. So pulleys, screws and gears are all mechanics because they can make things that go against nature, including automata like clocks. But after the emergence of the wheelwork clock in Europe around the end of the thirteenth century, this view of mechanism is completely reversed.[24] 'By 1377', Pierre Hadot tells us, 'Nicholas Oresme in his *Treatise on the Heavens and the World* was representing the motion of the heavens as that of a clock which, after being fashioned by God, continues to move by itself according to the laws of mechanics. This metaphor was to survive for centuries.'[25]

Far from being contrary to nature, the governing principles of nature came to be understood precisely as mechanical principles. Just as humans create clocks that run automatically, so God created nature that subsequently runs by mechanical law. To be sure, a purely mechanical world of microscopic cranks and levers might seem devoid of theological sense. But for the mechanists of the late medieval and early modern period, the mechanisation of nature was not only compatible with the Christian idea of a transcendent God, it actually demanded it, like a clock demands a clockmaker.[26]

For Descartes, the idea of nature as a clockwork machine is no mere metaphor – it is completely literal. In his *Principles*, Descartes asserts that he does not 'recognise any difference between artefacts and natural bodies'.[27] The only difference between human artifacts and naturally occurring bodies is their size:

> the operations of artefacts are for the most part performed by mechanisms which are large enough to be easily perceivable by the senses – as indeed must be the case if they are to be capable of being manufactured by human beings. The effects produced in nature, by contrast, almost always depend on structures which are so minute that they completely elude our senses ... it is no less natural for a clock constructed with this or that set of wheels to tell the time than it is for a tree which grew from this or that seed to produce the appropriate fruit.[28]

To be sure, Descartes seems to identify the natural and the artificial here. Human technologies like clocks, light bulbs and bicycles are as natural as the sun, lightning and horses. But Descartes identified nature and artifice here on the level of machines only to draw the distinction on another level entirely: by claiming that all corporeal and physical things were mechanical, he relegated true creative power to the mind alone. Machines may very well be natural, but now it is the modern subject who stands over and against nature. The mechanisation of nature thus allows for the relocation of the human subject outside of nature, as a driver that operates upon the machine, and the cogito asserts its dominion over the clock.

The extent of this dominion is perhaps most clearly visible in Descartes' view of animals, which he held to be soulless and insentient creatures, mere machines (*bête-machines* or 'animal machines'). While human life enjoys the benefit of mental substance that distinguishes it from the brute mechanism of nature, there is no cogito in nature: plants, animals and even the corporeal dimensions of the human body are purely physical, mechanical tools. This means that animals are machines in the exact same sense that clocks are machines. In the words of one commentator, Descartes' view of animals presents 'a grim foretaste of a mechanically minded age' that 'brutally violates the old kindly fellowship of living things'.[29] Similarly, Georges Canguilhem thinks that with Descartes

> We find ourselves in the presence of an attitude typical of Western man. The theoretical mechanization of life and the technical utilization of the animal are inseparable. Man can make himself master and possessor of nature only if he denies all natural purpose and can consider all of nature, including, apparently, animate nature – except for himself – to be a means.[30]

Not all commentators think that Descartes maintained a view that animals are exactly like machines (meaning that they do not feel pain or have sense-awareness).[31] But the mechanisation of nature is not likely to encourage an ecological view of the human within its environment. As previously mentioned, the bifurcation of nature into a physical aspect and a conscious aspect facilitates a picture of the human in confrontation with nature, rather than in an ecological relation, whether humans are conceived as stewards of the earth or human life is conceived as emergent within an environmental system. By reducing nature to a clock, the cogito can view its concrete environment as purely a means, an object there for technical utilisation. Consequently, Hadot finds that the dominion of humans over nature entailed by seventeenth-century mechanism is a profoundly

The Clock and the Cogito

Christian characteristic, echoing God's exhortation to Adam and Eve: 'Subjugate the earth'.[32]

Initially at least, it seems that seventeenth-century mechanism is inseparable from theology because the sheer conception of a machine demands a prime mover, a spontaneous and creative impulse that starts (or, as we shall see, sustains) the machine. Matter is thoroughly passive – it could never start itself – and so it requires the activity of mental substance, first in the form of a God to inaugurate the motion, and subsequently in the form of the cogito (made in the likeness of God, with the mark of the craftsman upon his work) to intervene in its motions. Consequently, mechanism as a science not only requires but also promotes the idea of God as the great engineer of nature, which at once guarantees the intelligibility of its design. But behind this inseparability between mechanism and theology is the notion that nature was created as a machine is created, that is, at a certain moment in time that could be identified by means of a clock. The most interesting aspect of the correspondence between seventeenth-century mechanism and theology lies in an obvious but easily overlooked dimension of the clock, because the clock is a machine, but it is a machine that measures time. In other words, the comprehension of nature in terms of a clock does not merely suggest that nature is a machine, but also suggests that nature was created and operates according to clock time, the time of machines.

THE TIME OF CLOCKS AND MECHANISM

A clock measures time through some periodic movement. Thus, a clock determines moments of time as a series of successive motions, which fits nicely with the doctrine of mechanism, understood as a sequence of movements. Machines, in other words, work by successive motions of parts, where one part moves another, and that yet another. But this means that the only difference between one moment and another is the arrangement of parts. If the entirety of corporeal substance is a machine of this sort – an immense plenum of interlocking cogs of various shapes and motions – then the only difference that distinguishes one moment from the next is the material configuration of natural parts. Time itself, apart from motion, is thereby rendered homogeneous, in the sense that there is no difference between the past and the present, the present and the future, or between one moment and the next. All of these 'times' are simply moments in a line of homogeneous time, one moment is no different from another, since time by itself merely expresses the succession of a sequence of moving parts. In this fashion, the history of nature is interpreted as the history of a series of material configurations passing through successive moments in time.

Both time and matter are abstractly understood in this clockwork picture of nature. Time is abstractly understood because it is comprised of homogeneous motions, and matter is abstractly understood because it is comprised of homogeneous stuff – there is no basic difference between one corporeal substance and another, but only superficial differences such as shape, size and motion (i.e. there is no basic difference between various primary qualities. This abstract picture of nature is well suited to mechanical physics, since primary qualities express precisely the measurable qualities of bodies, apart from the irreducibly qualitative features of bodies (the secondary qualities). Hence Descartes' project for a universal science or *mathesis universalis* becomes a real possibility only after the mechanisation of nature. As previously mentioned, the scholastics, who did not evaluate the subject and the world according to the epistemological standard of a certain and universal science, distributed nature quite differently, across a variety of qualities ranging from basic or 'primary' qualities such as the Hot and the Cold to spiritual qualities such as light or colour and even unknowable occult qualities such as magnetism.[33] In order for this distribution of nature to be reduced to two categories, the subject needed to become the transcendent cogito, and the corporeal needed to become the mechanistic clock.

In this sense, it comes as no surprise that the emergence of the mechanical clock in Europe precipitated the mechanisation of nature. The abstract idea of nature as a clockwork mechanism seems conditioned by the mechanical clock itself, since it is precisely that technical object which facilitates the abstract concept of time as a series of homogeneous motions. Throughout history, the concept of time has undergone broad changes in form: cyclical (life and death), eschatological (salvation) and linear (scientific). Within these broad changes, we find that different concepts of time and history were supported through different techniques and conventions for measuring and arranging time. So, for example, the ancient Greeks divided time into divine, human and natural time; the Middle Ages configured time in terms of the Easter cycle; the industrial period formulated labour time; and modern-day capitalism conceives of time in terms of interest and credit.[34] In each case, the techniques used to measure time,

The Clock and the Cogito

alongside the calendars and conventions that arrange time, inform a popular awareness of time, a time-sense. So a sundial provides a different sense of time from, say, an hourglass – in the former, we watch time arc toward the completion of the day, and in the latter time is literally rushing toward the end of a glass unit. While time had been contemplated in terms of abstract moments at least since antiquity, it had not, for that matter, been calculated abstractly until the mechanical clock, which meted out hours of equal movements.[35] It would be difficult to imagine the doctrine of mechanism arising in a social context where time was measured by horary prayers, sundials, candles, water-clocks, or any combination of such time-measuring techniques. In order to see nature as pure mechanism, one would have to see it in an artifact, a mechanical automaton that embodied the abstract view of time (homogeneous and successive motions) alongside the abstract view of matter (interlocking cogs of various shapes and sizes moving in sequences). Thus, the mechanical clock allowed its viewers to perceive a machine that operated according to the principles of mechanism, where at each moment, successive movements are generated by the contact of material parts. Furthermore, the intelligibility of the mechanical clock, engineered by humans, provided credence for the intelligibility of (mechanical) nature, engineered by God. In this sense, one can argue that the concept of nature as mechanical automatism is conditioned by the automatism of the mechanical clock.

Later chapters will interrogate the distinction between nature and technology, offering some discussion on how nature and time are conceived according to diverse technologies. For now, we may simply note that while we popularly consider technology as something opposed to nature (as man is opposed to or in confrontation with nature), we have nonetheless historically understood nature in terms of our technologies. Hadot, as previously mentioned, notes that the idea of nature as mechanism follows upon the invention of the mechanical clock, and Daniel Smith notes that other commentators have discussed the connection between major technological changes and changes in how we conceptualise nature.[36] In 1606, the Spanish inventor Jerónimo de Ayanz y Beaumont patented the steam engine, an energetic machine. By the beginning of the nineteenth century, the steam engine was powering locomotives. And while the characterisation of nature as a clock persisted for centuries after Nicholas Oresme's *Treatise on the Heavens and the World* in 1377, this view would eventually be supplanted by a characterisation of nature in terms of energy. The Cartesian laws of mechanism begin to erode with the energetic view of nature, visible already in Leibniz and Newton, but ultimately and most influentially expressed with the thermodynamics of Joule and Kelvin in the nineteenth century. And now, after the invention of the computer, we discuss nature in terms of information. The workings of nature, from the Higgs boson to DNA and RNA, are now described in terms of codes and messages. So after the production of the informational machine, the concept of nature becomes informational. In this sense, it becomes possible to argue for a *technological history of the concept of nature*, since we have always understood nature in terms of technology. While nature is typically considered as opposed to technology and artifice, such a history would reveal that the concept of nature is more fully understood as a result of events, practices and technologies – far from separating the human from nature in order to understand it, we may need to reinstall the human into nature precisely in order to understand it.

God, Mechanism and the Cogito

As previously mentioned, since the doctrine of mechanism demands that matter be completely passive, some primordial activity must be required in order to set the clock in motion, hence the necessity of a prime mover. However, the activity of the prime mover cannot be understood in terms of passive mechanism – the motion of the prime mover cannot itself be a clockwork motion. If the inaugural act amounted to cranking a mechanical clock into gear, merely imparting a motion into a sequence of subsequent motions, then one might rightly ask what put that first act into motion, and so on. This threat of infinite regress was a source of consternation for many philosophers from Aristotle to the scholastics. If one follows a chain of efficient causation or a chain of movement all the way back, then one must arrive at a first cause or a first mover which, by its very priority, must be unmoved or uncaused. Descartes characterises this first cause as self-caused or *causa sui*.[37] Behind this slightly different characterisation of the prime mover lies a crucial departure from the scholastic picture of the world toward the mechanistic view of nature.

The first of Aquinas's celebrated 'Five Ways' or proofs for God's existence is based on the necessity of a prime mover in order to get the world turning:

The Clock and the Cogito

> The first and more manifest way is the argument from motion. It is certain, and evident to our senses, that in the world some things are in motion. Now whatever is in motion is put in motion by another, for nothing can be in motion except it is in potentiality to that toward which it is in motion; whereas a thing moves inasmuch as it is in act. For motion is nothing else than the reduction of something from potentiality to actuality. But nothing can be reduced from potentiality to actuality, except by something in a state of actuality. Thus that which is actually hot, as fire, makes wood, which is potentially hot, to be actually hot, and thereby moves and changes it. Now it is not possible that the same thing should be at once in actuality and potentiality in the same respect, but only in different respects. For what is actually hot cannot simultaneously be potentially hot; but it is simultaneously potentially cold. It is therefore impossible that in the same respect and in the same way a thing should be both mover and moved, i.e. that it should move itself. Therefore, whatever is in motion must be put in motion by another. If that by which it is put in motion be itself put in motion, then this also must needs be put in motion by another, and that by another again. But this cannot go on to infinity, because then there would be no first mover, and, consequently, no other mover; seeing that subsequent movers move only inasmuch as they are put in motion by the first mover; as the staff moves only because it is put in motion by the hand [m.e.]. Therefore it is necessary to arrive at a first mover, put in motion by no other; and this everyone understands to be God.[38]

Descartes announces his departure from Aquinas's way of proceeding ('It is certain, and evident to our senses, that in the world some things are in motion') in the *Meditations* by claiming that he did not base his argument for God's existence 'on the fact that I observed there to be an order or succession of efficient causes among the objects perceived by the senses'.[39] In other words, he did not base his argument for God's existence on the movement of nature – God does not exist because the clock needs to be set in motion. The 'second proof' for God's existence in the Third Meditation is not drawn along the lines of motion and efficient causation, but rather along the distinction between the finite and the infinite. It is evident, Descartes reckons, from the very fact that the cogito doubts, that it must lack something. The cogito questions, doubts and hazards judgement precisely because it realises that it is not a full, self-sufficient whole, but rather a dependent, partial being. It is only in this sense that it can ask, not only the question of its origin, but any question whatsoever – because it knows that it does not know. Knowing that he is not his own author (he did not create himself), and that he does not have the power to preserve himself in time, he finds that something else must have created him and instilled within him the idea of a self-sufficient whole, some infinity, that is, by which he can recognise his finitude. How else would the cogito innately recognise that it is not a whole and self-sufficient being, unless there was some basic idea of a whole and self-sufficient being against which he could recognise his defect? In this fashion, Descartes finds that his recognition of his own finitude actually demands a native idea of infinity within the cogito itself, like a mark of the craftsman upon his work. The fact that he knows he is dependent entails the thought of something independent, which, by definition, is self-caused (i.e., it is not dependent on anything else, thus must cause and sustain itself, otherwise it would be dependent on something else). Consequently, everything that proceeds from this lack and finitude – all of his desires, all of his questions, in short, the entire pursuit of science and philosophy – is a calling of the cogito toward the thought of infinity, which is literally God drawing the cogito toward himself.

The cogito, along with the world, is dependent upon something independent (self-caused), which means that in so far as there are dependent substances, there is an independent substance that sustains those dependents. Thus, the characterisation of God as *causa sui* reconfigures the prime mover as an infinite cause, not only a cause 'at the beginning', so to speak, but as a continual cause. In other words, God is no longer merely an efficient cause or a prime mover. Descartes is forthright about his aim to reconfigure the prime mover in this fashion: 'the question I asked concerning myself was not what was the cause that originally produced me, but what is the cause that preserves me at present. In this way I aimed to escape the whole issue of the succession of causes.'[40] If Descartes had asked about the origin of nature or the origin of the cogito, he would have embarked upon a chain of successive causation. This would have been an inquiry about what came first – we have nature and the cogito as an effect, so what preceded that effect as a cause? But there is something interesting about God that places him over an above this kind of causation: 'the natural light does not establish that the concept of an efficient cause requires that it be prior in time to its effect. On the contrary, the concept of a cause is, strictly speaking, applicable only for as long as the cause is producing its effect, and so it is not prior to it.'[41] God must therefore be understood, not as a 'first' cause (in that sense he would be 'before' the clock and the cogito he effected), but rather as a simultaneous cause. God is not understood in terms of succession but in terms of infinite simultaneity. And it is through the figure of infinite simultaneity (or, what amounts to the same, eternity) that we are to understand God's independence:

> I do readily admit that there can exist something which possesses such great and inexhaustible power that it never required the assistance of anything else in order to exist in the first place, and does not now require any assistance for its preservation, so that it is, in a sense, its own cause; and I understand God to be such a being. Now I regard the divisions of time as being separable from each other, so that the fact that I now exist does not imply that I shall continue to exist in a little while unless there is a cause which, as it were, creates me afresh at each moment in time. Hence, even if I had existed from eternity, and thus nothing had existed prior to myself, I should have no hesitation in calling the cause which preserves me an 'efficient' cause. By the same token, although God has always existed, since it is he who in fact preserves himself, it seems not too inappropriate to call him 'the cause of himself'. It should however be noted that 'preservation' here must not be understood to be the kind of preservation that comes about by the positive influence of an efficient cause; all that is implied is that the essence of God is such that he must always exist.[42]

Time can be divided in such a way that each successive moment is completely separable from another, thus requiring a cause outside of time in order to sustain the continuity of succession. The eternity of God, then, is not on the same order of time as efficient causality, the successive time that obtains in the natural order of things. Succession is clock time, one moment after another, but the cause and preservation of succession is not itself successive. In this fashion, God does not share the same temporality of the clock, or even the cogito for that matter. The temporality of God is an eternity that sustains the continuity of the clock and the cogito, creating the world afresh at each moment in time. This becomes an axiom for Descartes: 'There is no relation of dependence between the present time and the immediately preceding time, and hence no less a cause is required to preserve something than is required to create it in the first place.'[43] Successive time is not causal, it is as passive as matter, and requires the creative spontaneity of God, not only for its beginning, but for its entire duration.

From the perspective of time, the clock and the cogito are in equal dependence upon God. Both require God's eternal causation for their creation and continuity, for their existence and duration. The clock, as previously mentioned, embodies a time of homogeneous and successive motions, which is precisely the temporality of mechanism understood as a series of passive material sequences. But while the cogito is neither material nor (completely) passive, the temporality underpinning the cogito remains much the same:

> For a lifespan can be divided into countless parts, each completely independent of the others, so that it does not follow from the fact that I existed a little while ago that I must exist now, unless there is some cause which as it were creates me afresh at this moment – that is, which preserves me. For it is quite clear to anyone who attentively considers the nature of time that the same power and action are needed to preserve anything at each individual moment of its duration as would be required to create that thing anew if it were not yet in existence. Hence the distinction between preservation and creation is only a conceptual one.[44]

Reiterating the infinite divisibility of successive time, Descartes finds that the lifetime of the cogito can be divided just as the sequences of matter can be divided.[45] The cogito, just like the clock, operates according to successive and homogeneous time. Consequently, the duration of the cogito, just like the duration of nature, the successive time of both, depends on God's spontaneous eternity. The distinction between creation, existence and duration thus reduces (as anyone who pays attention to the nature of time can see) in reality to a causal dependence upon the time meted out to the clock and the cogito by God. The triangular relation between God, mechanism and the cogito thus hinges upon a concept of time configured in terms of succession. But we at once see that time is understood in terms of mechanism throughout – active mental substance is distinct from passive corporeal substance, but the time underpinning the cogito is nonetheless mechanistic time. Thus, while Descartes maintains that the cogito is active by virtue of thinking, it actually acquires some passivity by virtue of its dependence upon mechanical time. This is the subject of Kant's well-known criticism of the Cartesian cogito – the 'I' of the 'I am' that is determined as existing by the 'I think' is not the same as the 'I' of the 'I think'. In other words, the 'I am' that is determined as existing in time is recognised only after thinking, as a phenomenal subject that is re-presented after the activity of thinking (thus requiring the succession of time for its recognition). As Deleuze writes, 'the spontaneity of which I am conscious in the "I think" cannot be understood as the attribute of a substantial and spontaneous being, but only as the affection of a passive self which experiences its own thought – its own intelligence, that by virtue of which it can say *I* – being exercised in it and upon it but not by it'.[46] Kant

The Clock and the Cogito

discovered that the form under which the 'I am' is determined is the form of time. By extracting all of time from the cogito and placing it in the causal efficacy of God, Descartes had rendered the subject of the cogito as passive as matter (which is why Kant would reinsert time within the cogito as its condition). The activity in the thinking cogito may be a small spark of spontaneity drawn in the likeness of God, but ultimately it reduces to an instantaneous spark requiring the continuity of God for its existence and duration. In this sense, the unity of the cogito (the unity of the 'I am' and the 'I think') is entrusted to the unity of God, who facilitates the duration of the cogito throughout successive time.

Thus, while the cogito is a mental substance, it is nonetheless a mental substance residing in a machine – a *cogito in rem mechanicam* – that unfolds in succession alongside the mechanical nature that it inhabits. We arrive, then, at a thoroughly spatial view of nature with Descartes, since even time is considered in terms of space (and motion in space). Like an extended line, the successive series of moments can be infinitely divided into parts. The time of the clock and the cogito is mechanistic time, a geometric view of time well-suited for the geometric view of matter in Descartes. The successive moments of time, like the material configurations that pass through them, are as homogeneous and infinitely divisible as geometric space itself. And it is this metaphysical picture of nature – as a series of material configurations passing through instants of time – that facilitates the seventeenth-century problem of error. Descartes had wondered how he could know, among other things, the objective reality beyond the confusion of his senses. This problem could only arise by splitting the world into an objective pole and a subjective pole, and Descartes aligns this bifurcation along the fault line of the cogito, separating intelligible geometrical qualities from sensory secondary qualities. The subject is thus placed in confrontation with the object, and can ask how he is to access, with certainty, the real qualities of the physical world beyond the forbidding requirement of going through his senses. The answer is that through God's veracity (manifest by the natural light), the real qualities of objects are geometrically and mechanistically intelligible, affording the cogito speculative access to the physical world. To be sure, this question of access can only be posed if the subject has been separated from, and placed in confrontation with, nature. But at once we notice that this metaphysical view of nature is traced along the lines of a perceptual picture of nature – a snapshot taken at an instant. And like a snapshot, this picture of nature requires us to conceptually pause time: it is only by pausing time that we obtain a picture of a subject confronting the objective substance of nature, inquiring about her mode of access to what is separate from her. It is here that we notice that the epistemological problem of certainty is facilitated by a metaphysical configuration of time based upon space: time must be a series of instants of material configurations – time is a flipbook of material snapshots. Mechanistic time, clock time, is not only at the bottom of the problem of error, it is at the bottom of the bifurcation of nature within which the problem is framed. The cogito has been separated from nature by configuring time in terms of space.

Characterising the isolation of Descartes' cogito as a 'brain in a vat', Latour notes that this model of certainty 'was not needed when the brain (or the mind) was firmly attached to its body and the body thoroughly involved in its normal ecology'.[47] Yet the bifurcation of nature and the problem of error would continue, starting with the empiricists after Descartes, who

> wondered whether the world could directly send us enough information to produce a stable image of itself in our minds. But in asking this question the empiricists kept going along the same path . . . God was out, to be sure, but the tabula rasa of the empiricists was as disconnected as the mind in Descartes's times . . . Since the empiricists' associative neural net was unable to offer clear pictures of the lost world, this must prove, they [the Kantians] said, that the mind (still in a vat) extracts from itself everything it needs to form shapes and stories. Everything, that is, except the reality itself.[48]

We know where that story ends. Reality is distributed into the phenomenal world that we can have knowledge of and the noumenal world that is necessary but difficult to acknowledge.[49] Unfortunately, we have inherited this concept of nature, which is structured by a dilemma: what we aspire to know is precisely what the conditions of knowledge preclude us from knowing.[50] Foucault often notes that we must separate ourselves from something in order to know it. If separation, or at least differentiation, is a condition of knowledge, then perhaps we can only know nature by objectifying it. Yet the history of philosophy offers some alternatives to the bifurcated picture of nature and, crucially, these alternatives involve a reconfiguration of the concept of time. We have seen that successive time is a condition of bifurcation, as one must pause time at an instant in order to separate a subject from the objective world that confronts it, thereby begging the epistemological question of access to the objective.

Presumably, then, if one is to think beyond bifurcation, one must first reconfigure the concept of time that subtends it, which in turn will reconfigure the role of knowledge vis-à-vis nature.

Notes

1. René Descartes, *Selected Philosophical Writings*, trans. J. Cottingham, R. Stoothoff, D. Murdoch (Cambridge: Cambridge University Press, 1988), p. 76. Compare with Descartes' *Discourse*: 'seeing that [philosophy] has been cultivated for many centuries by the most excellent minds and yet there is still no point in it which is not disputed and hence doubtful, I was not so presumptuous as to hope to achieve any more in it than others had done. And, considering how many diverse opinions learned men may maintain on a single question – *even though it is impossible for more than one to be true* – I held as well-nigh false everything that was *merely* probable' (*Selected Philosophical Writings*, p. 23).
2. Gilles Deleuze and Claire Parnet, 'I as in Idea', in *From A to Z*, Subtitles Charles Stivale (Los Angeles: Semiotext(e), 2012), Disc Two.
3. Hadot, *The Veil of Isis*, pp. 124–5.
4. Descartes, *Selected Philosophical Writings*, p. 81.
5. Descartes, *Selected Philosophical Writings*, p. 83.
6. Descartes, *Selected Philosophical Writings*, p. 82.
7. Descartes, *Selected Philosophical Writings*, pp. 83–4.
8. Descartes, *Selected Philosophical Writings*, pp. 84–5.
9. Descartes, *Selected Philosophical Writings*, p. 85.
10. Descartes, *Selected Philosophical Writings*, p. 111.
11. For an excellent analysis of the imagination's role in mathematical reasoning, particularly in terms of abstract numbers, see the conclusion to Stephen Gaukroger's essay, 'The Nature of Abstract Reasoning: Philosophical Aspects of Descartes' Work in Algebra', in *The Cambridge Companion to Descartes*, ed. John Cottingham (Cambridge: Cambridge University Press, 1992), pp. 91–114.
12. For Descartes, imagination presents only confused representations, a distinction that can be witnessed when it is compared with the power of understanding: 'I notice quite clearly that imagination requires a peculiar effort of mind which is not required for understanding' (*Selected Philosophical Writings*, p. 111). Descartes' critique of imagination and validation of understanding here have a temporal underpinning: while the imagination can momentarily hold a picture (say, of a brick wall) in the mind, it cannot achieve the stability of judgement and understanding that would allow one to affirm and count the bricks. The imagination attempts to stabilise the image of a brick wall, but cannot afford a clear and distinct idea of the number of bricks because the imagination admits of duration.

13. Descartes, *Selected Philosophical Writings*, p. 89; italics mine.
14. Descartes, *Principles of Philosophy*, I, 68, in *The Philosophical Writings of Descartes, Vol. 1*, p. 217.
15. Descartes, *Principles of Philosophy*, I, 71, in *The Philosophical Writings of Descartes, Vol. 1*, p. 219.
16. In an excellent article, Lisa Downing finds that, while Descartes 'has an argument that he takes to show a priori that sensible qualities cannot be attributed to the material world', that argument ultimately fails, 'leaving him with at best a partly empirical case for removing the sensible qualities, based on the purported explanatory success of his physics'. Lisa Downing, 'Sensible Qualities and Material Bodies in Descartes and Boyle', in *Primary and Secondary Qualities: The Historical and Ongoing Debate* (Oxford: Oxford University Press, 2011).
17. On the theological roots of Cartesian physics, see chapter 3 of Daniel Garber, *Descartes' Metaphysical Physics* (Chicago: University of Chicago Press, 1992), pp. 63–93.
18. Descartes, *The World or Treatise on Light*, in *The Philosophical Writings of Descartes, Vol. 1*, p. 83.
19. Descartes, *Principles of Philosophy*, IV, 198; in *The Philosophical Writings of Descartes, Vol. 1*, pp. 284–5.
20. See Lisa Shapiro, 'Princess Elizabeth and Descartes: The Union of the Soul and Body and the Practice of Philosophy', *British Journal for the History of Philosophy*, Vol. 7, No. 3 (1999), pp. 503–20.
21. Descartes, *Selected Philosophical Writings*, pp. 229–30.
22. Descartes, *Selected Philosophical Writings*, p. 231.
23. T. Willis, 'The Anatomy of the Brain and the Description and Use of the Nerves', in *The Remaining Medical Works of That Famous and Renowned Physician Dr. Thomas Willis*, trans. S. Pordage (London, 1681). See Gert-Jan Lokhorst, 'Descartes and the Pineal Gland', *The Stanford Encyclopedia of Philosophy*, ed. Edward N. Zalta, Summer 2011 edition.
24. China had enjoyed the escapement mechanism (which regulated the water-powered wheel driving the clock) before Europe (around 725 ad), and Joseph Needham finds that China is the missing link between early water clocks and the later mechanical clocks in the West. See Joseph Needham, *Heavenly Clockwork: The Great Astronomical Clocks of Medieval China* (Cambridge: Cambridge University Press, 1986).
25. Hadot, *Veil of Isis*, p. 85. Also in Arno Borst, *The Ordering of Time: From the Ancient Computus to the Modern Computer* (Chicago: University of Chicago Press, 1993), p. 97.
26. Hadot cites Voltaire's well-known lines: 'The universe embarrasses me, and I cannot imagine that such a clock should exist without a clockmaker.' Hadot, *The Veil of Isis*, p. 127.

27. Descartes, *Principles of Philosophy* (IV, 203), p. 288.
28. Descartes, *Principles of Philosophy* (IV, 203), p. 288.
29. Alexander Boyce Gibson, *The Philosophy of Descartes* (London: Methuen, 1932), p. 214.
30. Georges Canguilhem, *Knowledge of Life* (New York: Fordham University Press, 2008), p. 84.
31. John Cottingham, 'A Brut to the Brutes? Descartes' Treatment of Animals', *Philosophy*, Vol. 53 (1978), pp. 551–9.
32. Hadot, *Veil of Isis*, pp. 129–30.
33. Pasnau, 'Scholastic Qualities, Primary and Secondary', p. 43.
34. There are many excellent books on the history of the concept of time and the technologies and conventions for understanding time. See Stephen Toulmin and June Goodfield, *The Discovery of Time* (New York: Harper & Row, 1965). Also see Carlo Cipolla, *Clocks and Culture: 1300–1700* (New York: W. W. Norton & Company, 1978), and Borst, *The Ordering of Time*.
35. Arno Borst, who finds the revolutionary influence of the mechanical clock overrated by scholars, nonetheless acknowledges that it significantly changed the social life of Europe. Work was increasingly timed by clocks (and not by tasks), and 'the clock was a symbol for a measured way of life in the midst of chaotic circumstances'. Borst, *The Ordering of Time*, p. 95.
36. Daniel Smith, 'Deleuze on Technology and Thought', unpublished paper.
37. Jean-Luc Marion develops a thesis that the concept of *causa sui* is a Cartesian advancement beyond scholastic thought, which typically considered God as an 'unmoved' mover or an 'uncaused' cause. Jean-Luc Marion, *On the Ego and on God: Further Cartesian Questions*, trans. Christina M. Gschwandtner (New York: Fordham University Press, 2007), pp. 140–4.
38. Aquinas, *Summa Theologica* I, q. 2, a. 3, response in <http://www.newadvent.org/summa>.
39. Descartes, Objections and Replies, in *The Philosophical Writings of Descartes, Vol. 2*, p. 77.
40. Descartes, Objections and Replies, in *The Philosophical Writings of Descartes, Vol. 2*, p. 77.
41. Descartes, Objections and Replies, in *The Philosophical Writings of Descartes, Vol. 2*, p. 78.
42. Descartes, Objections and Replies, in *The Philosophical Writings of Descartes, Vol. 2*, pp. 78–9.
43. Descartes, Objections and Replies, in *The Philosophical Writings of Descartes, Vol. 2*, p. 116.
44. Descartes, Objections and Replies, in *The Philosophical Writings of Descartes, Vol. 2*, p. 33. The standard view of time and causality in

Descartes maintains that he understood time in terms of discontinuous, independent instants. See, for example, Jean Wahl, *Du Rôle de l'Idée de l'Instant dans la Philosophie de Descartes* (Paris: Alcan, 1920). Secada, however, argues that, contrary to the standard view, 'as far as we can know, Descartes had no views as to the continuity or discontinuity of time'. J. E. K. Secada, 'Descartes on Time and Causality', *The Philosophical Review*, Vol. XCIX, No. 1 (January 1990), pp. 45–72.

45. The Latin version of 1641 says a lifetime can be divided (*dividi potest*) into 'innumerable' parts: 'Quoniam enim omne tempus vitae in [49] partes innumeras dividi potest.' But the Duc de Luynes French translation in 1647 (with the supervision of Descartes) uses 'infinité': 'Car tout le temps de ma vie peut être divisé en une infinité de parties.'
46. Gilles Deleuze, *Difference and Repetition*, trans. Paul Patton (New York: Columbia University Press, 1994), p. 86.
47. Bruno Latour, *Pandora's Hope: Essays on the Reality of Science Studies* (Cambridge, MA: Harvard University Press, 1999), p. 4.
48. Latour, *Pandora's Hope*, pp. 4–5. Whitehead brilliantly notes, 'Induction presupposes metaphysics. In other words, it rests upon an antecedent rationalism. You cannot have a rational justification for your appeal to history till your metaphysics has assured you that there is a history to appeal to; and likewise your conjectures as to the future presuppose some basis of knowledge that there is a future already subjected to some determinations.' Whitehead, *Science and the Modern World*, p. 44. The concept of induction, which operates within a confrontational model of epistemology, itself relies on a metaphysical presupposition that we are subjects in an objective flux of time and space.
49. Jacobi's well-known criticism of the critical philosophy is that, 'Without the presupposition [of the "thing in itself,"] I was unable to enter into [Kant's] system, but with it I was unable to stay within it.' F. H. Jacobi, *Main Philosophical Writings and the Novel Allwill* (Montreal: McGill-Queens University Press, 2009).
50. The bifurcation of nature, even in its Cartesian formulation, persists in contemporary philosophy. Quentin Meillassoux's *After Finitude* restates the Cartesian problem of certainty in basically the same form. Frustrated with the centrality of the subject in the last two hundred years of philosophy since Kant, Meillassoux's book begins with a bold call for a return to pre-critical philosophy: 'the theory of primary and secondary qualities seems to belong to an irremediably obsolete philosophical past. It is time it was rehabilitated.' Quentin Meillassoux, *After Finitude: An Essay on the Necessity of Contingency* (London: Continuum, 2008), p. 1. The Cartesian dimension of Meillassoux's project has been noted by many commentators. Brassier, for example, writes, 'strange as it may seem, Meillassoux's project is essentially Cartesian: by rehabilitating thought's access to the absolute, he

hopes to demonstrate mathematical science's direct purchase on things in-themselves'. Ray Brassier, *Nihil Unbound: Enlightenment and Extinction* (New York: Palgrave Macmillan, 2010), p. 69. Ennis makes essentially the same claim: 'Meillassoux is sympathetic to the pre-critical commitment to primary qualities – especially the Cartesian claim that primary qualities are mathematical in nature and thereby accessible.' Paul Ennis, *Continental Realism* (Washington: Zero Books, 2011), p. 5. And Meillassoux himself expressly states that he aims to rediscover 'an in-itself that is Cartesian, and no longer just Kantian . . . to legitimate the absolute bearing of the mathematical'. Meillassoux, *After Finitude*, p. 111. The most basic difference is that Meillassoux attempts to resolve the problem of error without God's veracity. Yet by rehabilitating the Cartesian bifurcation of nature, he also incurs the burden of explaining how time passes without God's continual creation. The fact that Meillassoux configures his concept of nature in terms of clock time is evident from the way he poses the problem of post-Kantian philosophy. He claims that the problem with philosophy is its inadequate comprehension of *ancestrality*, which is the scientific claim to knowledge about the world that pre-dates consciousness. The figure of ancestrality is the *arche-fossil*, an entity that designates, through scientific methods of detection (isotopic rates of radioactive decay, the luminescent emissions of stars), a reality that precedes consciousness. Now, Meillassoux fashions his argument in a way that this designated reality is not simply spatio-temporally anterior to human consciousness, rather, it is a reality that predates the spatio-temporal *conditions* of representation and knowledge itself, and so it is a reality flaunted as devastating evidence against critical philosophy: '*science can think a world wherein spatio-temporal givenness itself came into being within a time and a space which preceded every variety of givenness*'. Meillassoux, *After Finitude*, p. 22. The clock, in other words, was operative and material configurations were enduring through successive instants long before the cogito was inserted into the machine. The past is thus considered as a dated present, a yellowing snapshot, a faded material configuration, in short, an object with knowable primary qualities. Consequently, Meillassoux encounters considerable difficulties when discussing time, duration, events and affects.

2

The Polarisation of Nature

It would be hard to overstate the significance of the Cartesian bifurcation of nature into primary and secondary qualities, for three main reasons. First, the bifurcation enables a conception of nature that abstracts entities from their concrete environments. As previously mentioned, for Aristotle, bodies are tied essentially to a place – a rock falls down to the earth and a flame rises to the sky because those bodies naturally belong in those places. For Newton, on the other hand, bodies are abstracted from place – a body's motion is no longer related to the sky or the earth, the only thing that matters are their positions relative to other bodies. As Heidegger notes, with Newton, 'nature is no longer an inner capacity of a body, determining its form of motion and its place. Nature is now in the realm of the uniform space-time context of motion, which is outlined in the axiomatic project and in which alone bodies can be bodies as a part of it an anchored in it.'[1] With regard to this axiomatisation of nature, one of the crucial figures between Aristotle and Newton is Descartes, since he explicitly abstracts bodies from their concrete environments by conceiving of nature in terms of primary or geometrical qualities, making all bodies a part of *res extensa*, distinguished only in terms of shape, size, motion, etc. The idea that space remains the same throughout earth, mountains, sky and heavens, as an abstract coordinate grid subtending the relations of bodies within that grid, is one of Descartes' contributions to the philosophy of nature, and it is the major ramification of the bifurcation for the modern scientific interrogation of nature.

The second ramification of the bifurcation is institutional and disciplinary. As previously mentioned, the bifurcation separates the human from nature and places him in confrontation with it. While the human body is partly a body like any other, subject to the same laws of mechanism and understood abstractly in terms of geometrical qualities, it nonetheless is somehow united to a soul, a purely mental substance that is independent of *res extensa*. The conscious soul is thus the distinguishing characteristic of the human, granting man an activity that constitutes his dominance over the passivity of nature. Ultimately, however, this separation of the conscious subject from the material body facilitates the broad disciplinary distinction between the natural sciences and the social sciences and humanities. After all, it is only by bifurcating nature that one can separate sciences into those that deal with the natural world and physical bodies, on the one hand, and those that deal with humans and society, on the other.

The third significant ramification of the bifurcation is theoretical and philosophical. Splitting nature into a material dimension and a mental dimension sets the stage for one of the most vigorously contested debates in the history of philosophy: the confrontation between materialist realism and idealism. Much of natural philosophy after the seventeenth century can be characterised as an attempt to negotiate the relation between mind and matter.[2] Deleuze recounts the impasse between consciousness and extended space as follows:

> In consciousness there would only be images – these were qualitative and without extension. In space there would only be movements – these were extended and quantitative. But how is it possible to pass from one order to the other? How is it possible to explain that movements, all of a sudden, produce an image – as in perception – or that the image produces a movement – as in voluntary action? . . . What appeared finally to be a dead end was the confrontation of materialism and idealism, the one wishing to reconstitute the order of consciousness with pure material movements, the other the order of the universe with pure images in consciousness.[3]

Several philosophers have attempted to overcome the bifurcation. Whitehead and Bergson's critical appraisals of the bifurcation have already been mentioned, and they both made somewhat similar attempts at developing alternative philosophies of nature. This chapter will examine Bergson's reformulation of the mind and matter problem. Bergson offers a remarkable counterpoint to Cartesian formulations of nature because, instead of separating the mind from matter in terms of space, he places both the mind and matter upon a continuum of time. As we shall see, this reformulation of the problem draws a line perpendicular to the opposing end points of the debate between realism and idealism, and it affords a new perspective on the relation between science and philosophy.

The Metaphysics of Perception: Image, Matter and Memory

'It would greatly astonish a man unaware of the speculations of philosophy', Bergson writes, 'if we told him that the object before him, which he sees and touches, exists only in his mind . . . [and] we should astonish him quite as much by telling him that the object is entirely different from that which is perceived in it, that it has neither the colour ascribed to it by the eye nor the resistance found in it by the hand.'[4] Since common sense suggests that the objects in the world really are in the world, existing independently from us, Bergson starts from a common-sense perspective in order to bypass the pitfalls of both idealism and realism. Whereas Descartes 'put matter too far from us when he made it one with geometrical extensity', Berkeley put it too close to us by placing it within our minds and making it into a pure idea.[5] Both realism and idealism separate mind and matter and then try to negotiate some correspondence between the two terms. Bergson, on the other hand, will navigate between realism and idealism by deferring the separation of mind and matter, instead focusing on perception as it is commonly understood.

From the perspective of common sense, the world is full of images – a dog barking, a tram rolling by, office chairs, a rainy day. These images are somewhat regular, in the sense that they have patterned and regular interactions – only rarely, if ever, does a dog turn into an office or a tram. As images, then, all things seem to act within their capacities, and what we normally call 'laws of nature' seem to hold sway.[6] However, there is one image that, while embedded among the rest, nonetheless seems to enjoy some amount of freedom from these laws of nature, as if it had some sort of privilege. This is the image of our body, and its apparent privilege arises from the fact that we not only perceive it, but we also feel it as a kind of centre. Our body-image functions as a kind of locus with regard to other images, in the sense that as we move closer or farther away from other images their qualities (odour, sound, visibility, etc.) increase or diminish. This is simply to say that we know when our food is burning, or when there are people at the door, and we conform to the arrangement of the furniture in our homes. Bergson simply accedes the apparent: 'My body is, then, in the aggregate of the material world, an image which acts like other images, receiving and giving back movement, with, perhaps, this difference only, that my body appears to choose, within certain limits, the manner in which it shall restore what it receives.'[7]

Unlike external, inanimate material images, the body-image is not automatic, and can decide upon its action within certain limits. This is why Bergson will ultimately identify the body together with its conscious involvement as a 'zone of indetermination' in the midst of image-movements that follow an automatically determined order.

The Polarisation of Nature

But what is initially noteworthy in Bergson's analysis of the body is that instead of engaging a hasty separation between the objective and the subjective, he decides to analyse perception exclusively in terms of images. And while he acknowledges a peculiarity of the body-image, he refrains from ascribing to it a power of representation, or image-production. Navigating an intuitive course that remains on the level of images is precisely what allows Bergson to avoid the trappings of both realism and idealism. The term 'image' itself operates between those two philosophies: 'by "image" we mean a certain existence which is more than that which the idealist calls a *representation*, but less than that which the realist calls a *thing* – an existence placed halfway between the "thing" and the "representation"'.[8] Since both the realist and the idealist separate the mind from matter, the halfway mark between a representation and a thing turns out to be right where the images appear to be – namely, wherever they are perceived within and without us. Thus, Bergson's intuitive or common-sense perspective rehabilitates the image beyond the representational framework that would render it an appearance of some reality behind it. The image is certainly apparent, but the apparent loses its philosophical sense because it is no longer distinguished from a supposed reality behind it.[9]

The Abstract Mechanics of Perception

While the body-image is a focal point among other images, it is nonetheless simply an image. By making the body an image like any other, Bergson refrains from ascribing a power of representation to it. But how, then, does representation occur? In other words, 'how could my body in general, and my nervous system in particular, beget the whole or a part of my representation of the universe?'[10] Again, while the body is a privileged centre (in the sense that it is not only perceptive but affective), that privilege does not afford it the ability to produce or represent other images. The image of our body, as an image, cannot produce the aggregate of images we call the universe, in the same sense that an image of a martini, or a plate of spaghetti, cannot produce a representation of the universe. If we take Bergson's identification of the body as basically an image among others (although a kind of compass among images), then we are led to conclude with him that our body, as *'an object destined to move other objects, is, then, a center of action; it cannot give birth to a representation'*.[11] Far from accounting for representation by reference to an alchemist mind which translates primary physical movement through neurons in order to produce the secondary qualities of conscious life, Bergson's theory of perception adds nothing to matter: '*I call* matter *the aggregate of images, and* perception of matter *these same images referred to the eventual action of one particular image, my body.*'[12]

To be sure, this is an action-oriented, machinic account of perception. If we were to construct an automaton, replete with auto-focusing lenses like those of a camera, heat-sensors studding its external casing, filters to discern the air surrounding it, levels to gauge its balance, and mobilising mechanisms to organise its movement, all in the service of a program designed to interact within an environment that provided the kind of action it was engineered to receive, we would not be far from Bergson's abstract understanding of perception, which he calls *pure perception*; in fact, we would have identified it perfectly. In order to highlight the action-oriented nature of perception, Bergson conjures a picture of a patient undergoing surgery that unplugs all the nerves in charge of transmitting the current of sensory movement from the periphery of the body to the spinal cord and brain:

> A few cuts with the scalpel have severed a few bundles of fibers: the rest of the universe, and even the rest of my body, remain what they were before ... As a matter of fact, my perception has entirely vanished ... Sectioning of the centripetal nerves can, therefore, produce only one intelligible effect: that is, to interrupt the current which goes from the periphery to the periphery by way of the center, and, consequently, to make it impossible for my body to extract, from among all the things which surround it, the quantity and quality of movement necessary in order to act upon them. Here is something that concerns action, and action alone.[13]

How would a body that did not enjoy perception move? It would not have any reason to move one way over another, and it would in fact have little reason to move at all; there is, quite simply, no perception to act upon and hence no action. The world changed but little (only a minor snipping of a few nerves), yet the perception of the world has vanished entirely, leaving the body-image exactly how it always was, with only one difference: it can no longer discern the real and possible action of other images with regard to itself, and thus remains an insensitive patient of action.

Bergson finds it curious that, instead of considering the body (and the perceptual organs within it – including the brain) as material images just like any other material images, one might be inclined

The Polarisation of Nature

to impute some illustrative potential to one organ or another. As if nervous fibres in concert with the grey tissue of the brain could effect the entire representation of the universe. This would make the body-image into something not quite an image, rendering matter something not quite material. Perception, for Bergson, is only about matter, which is to say, images, since there is nothing different in kind among different material configurations. In matter there are only differences of degree, and images are nothing beyond images:

> But the truth is that the movements of matter are very clear, regarded as images, and that there is no need to look in movement for anything more than what we see in it . . . images themselves cannot create images; but they indicate at each moment, like a compass that is being moved about, the position of a certain given image, my body, in relation to the surrounding images.[14]

For Bergson, the body is an image-navigating machine, not an artistic fabricator who illustrates the cosmos. Cerebral disturbances are very much images within an aggregate of other images in the material world and, while they form part of the aggregate of images, they are not the genetic source of represented images. Adopting this 'intuitive' position we are led to the surprising conclusion that perception is basically a measure of action – perception is not produced by cerebral activity but rather is a relation of material images just like any other relation of images.[15] The brain's role in perception is just like the reflex functions of the spinal cord: 'There is, then, only a difference of degree – there can be no difference in kind – between what is called the perceptive faculty of the brain and the reflex functions of the spinal cord.'[16] Bergson admits that there is clearly some 'close connection between a state of consciousness (mind) and the brain (matter) . . . but there is also a close connection between a coat and the nail on which it hangs, for, if the nail is pulled out, the coat falls to the ground. Shall we say, then', he asks, 'that the shape of the nail gives us the shape of the coat, or in any way corresponds to it?'[17] Philosophers habitually discuss perception in terms of some correspondence between mind and matter, but thus far Bergson's intuitive method has avoided discussing such a correspondence, simply by avoiding the separation of mind from matter in the first place. For Bergson, pure perception is the calibrated movement of a body-image with regard to the images that constitute its surrounding environment. And this is precisely what intuition would suggest. In fact, if one were to eliminate philosophical speculation and simply attend to deliverances of the senses, one would find that all of the images perceived have one thing in common: images all tell us about our body's position with regard to them. A dog barking, the walls of a room, the pressure of the floor and chair, the hum of a tram – all these images reveal is our possible action within our local environment. The question of knowledge, if and when it is raised, does not arise from perceptual activity taken simply.

Thus, the method of intuition allows Bergson to draw a stunning conclusion: perception is not oriented toward knowledge (it is not designed to produce knowledge of some objective world), but rather toward action. Far from suggesting itself as an illustrator of representations that attempt, faithfully or not, to depict material objects, the perceptive body actually presents itself as a compass with which to navigate through the world of other images. Whereas Descartes considered images, whether secondary or primary, in terms of knowledge, Bergson considers images in terms of actual or possible action. The question is no longer how a mental subject can know a material object, since that question can only be asked once nature has been bifurcated. There is indeed a distinction between the subject and the object, but it will be drawn along different lines. The primary question for Bergson is, rather,

> *How is it that the same images can belong at the same time to two different systems: one in which each image varies for itself and in the well-defined measure that it is patient of the real action of surrounding images; and another in which all images change for a single image and in the varying measure that they reflect the eventual action of this privileged image?*[18]

This formulation of the distinction between the subject and the object bypasses the pitfalls of both realism and idealism – instead of representational images in the mind contending with material images in space, there is only one set of images for Bergson. The question becomes how a single plane of images can interact in two different systems; one system (the subjective or conscious system) selects images in terms of possible or eventual action, which is to say that it can take time to decide how to act upon other images, and the other system (the objective or material system) is simply automatic, responding in a determinate way to its surroundings. Conversely, the realist began with images outside of the subject and subsequently attempted to explain how perception could arise from a certain image called the brain, as if second-order 'representational

The Polarisation of Nature

images' could arise from first-order 'material images'. The idealist, on the other hand, began with interior states of consciousness only to be subsequently forced into providing an account of what was excluded from the outset, namely, the objective order of nature. Starting with the subject, the idealist attempts to work her way outside the mind, all the way down to the distant and foreign order of matter. Starting with the order of material nature, the realist attempts to work her way up to the baffling order of consciousness along with its perceptual illusions. But both realism and idealism share a representational framework that evaluates the image in terms of knowledge and speculation. As Bergson notes, for the realist and the idealist,

> *Perception has a wholly speculative interest; it is pure knowledge.* The whole discussion turns upon the importance to be attributed to this knowledge as compared with *scientific* knowledge. The one doctrine starts from the order required by science, and sees in perception only a confused and provisional science. The other puts perception in the first place, erects it into an absolute, and then holds science to be a symbolic expression of the real. But, for both parties, to perceive means above all to know.[19]

For Bergson, on the other hand, perception is not speculative but practical in nature: 'perception as a whole has its true and final explanation in the tendency of the body to movement'.[20] Again, perception is simply a compass that guides the possible action of our body-image in relation to the environment of images that surround it.

To be sure, if perception is apprehended actively instead of epistemologically, there is no difference between the image and matter. 'Matter', in Bergson's view, is nothing but the 'aggregate of "images"'.[21] There is no ghost or hidden power in matter that the image occludes or evinces; on the contrary, matter is taken as the image *simpliciter*. What may have appeared as a philosophically naive position thus becomes a philosophically provocative one – matter is not physical movement that we represent through images, rather, physical movement is the material image itself. There is no distinction, then, between matter, image and movement – as Deleuze writes, 'IMAGE = MOVEMENT'.[22] Again, there is no 'thing' behind the image that moves, but my body, the neural disturbances within it, and the environment without it, are all images identified by the actions and reactions that delimit them. And to be sure, this applies to all images in so far as they are images – even 'the atom', Deleuze notes, 'is an image which extends to the point which its actions and reactions extend'.²³

But what are we to say about dreams, hallucinations and the other threats that Descartes attempted to dispel from his project of certainty? How can Bergson maintain that images are material movements when we are constantly mistaken about images? The short answer to this question is that the question itself is badly stated – perceptual images are constituted by real movement and so they are neither right nor wrong, which in no way entails that events will unfold exactly how we anticipate them to unfold. But the longer answer requires an examination of the role of memory in perception, since it has been suppressed in the above analysis of pure perception.

The Concrete Metaphysics of Perception

In a case-study by the renowned neurologist Oliver Sacks, a man named Virgil, who had been functionally blind since the age of six, undergoes a type of eye surgery that restored his vision to a 'respectable 20/80'.²⁴ Or rather, it restored his ocular apparatus to a respectable degree of functionality, because after the surgery, Virgil could not really 'see', in any recognisable sense of the term. He could perceive light, as well as some movement, but it was all a meaningless blur.

William Molyneux, a seventeenth-century philosopher whose wife was blind, once posed a question to John Locke: 'Suppose a man born blind, and now adult, and taught by his touch to distinguish between a cube and a sphere [be] made to see: [could he now] by his sight, before he touched them ... distinguish and tell which was the globe and which the cube?'²⁵ Virgil certainly could not: 'On the day he returned home after the bandages were removed, his house and its contents were unintelligible to him, and he had to be led up the garden path, led through the house, led into each room, and introduced to each chair.'²⁶ A lifetime of habituation to a dark but elaborately tactile world prevented Virgil from settling into and navigating a world of sight. He had developed a certain technique for manoeuvring throughout his non-visual environment of images, and the bombardment of visible light had a paradoxically blinding effect. In fact, his entire identity had been wrapped up in his non-visual world, and his abrupt dislodgement from that world caused him considerable anxiety and exhaustion. However, after a while he made some advancements into the visual world – he began to 'see', and he was able to correlate newly visual images with his familiar non-visual memories. Clear vision would sometimes follow upon the heels of touch: on a visit to the zoo, he was pointed to a gorilla in the great-ape enclosure. Initially, however, Virgil thought that he was looking at a large man, and could not discern anything different about the gorilla except that it moved a little differently. But once he touched a life-size bronze statue of a gorilla nearby, he was able to 'see' the ape and accurately describe its posture, 'the way the knuckles touched the ground, the little bandy legs, the great canines, the huge ridge on the head, pointing to each feature as he did so'.[27] But the most fascinating aspect of the case of Virgil is that, after these small advances into the world of sight, he would be thrust back into his old world by strong emotional associations, like a visit from his family. His wife reported that once his family arrived on their wedding day, his sight seemed to 'have gone on vacation ... It is as though he has gone back to being blind!' And once his family said goodbye, his 'sight began clearing up right after they left'.[28] Seeing, then, is not simply a mechanical matter of ocular sensation. Sacks reports that 'it was very difficult, at times, to know what was going on, to distinguish between the "physiological" and the "psychological"'.[29]

In the previous section we analysed images exclusively in terms of movement and matter. But the memorial past is composed of neither movement nor matter – movements, after all, are always taking place in the present moment. We do not perceive the past and it is impossible to act upon past images; action and perception are exclusive to the time at hand, the time that is present at this moment. However, concretely, 'there is no perception which is not full of memories. With the immediate and present data of our senses, we mingle a thousand details out of our past experience. In most cases these memories supplant our actual perceptions, of which we then retain only a few hints, thus using them merely as "signs" that recall to us former images.'[30] When I see Werner, I am not operating purely on the register of the present image of Werner, but that image, rather, is imbued with an entire history that facilitates my quick recognition of Werner. In fact, if I were to eliminate memory and attend to the pure perception of Werner, I would never arrive at Werner because I would be lost in the confusion of image-movements, unable to focus, as it were, on any one of them – I would therefore share the experience of Virgil, without the habits and techniques that facilitate visual perception.[31] Thus, while images are purely movements, memory nonetheless plays an essential role with regard to how movements are arranged or selected. In fact, if there were no memory of the past, we would not, strictly speaking, perceive anything, since our actions would amount to purely automatic responses to external movements, much like the automaton of pure perception described above. We would not select any aspect of the movement as more or less relevant to our action, and neither would we focus, synthesise or 'contract' the series of movements into a coherent picture of our environment. One of Bergson's commentators explains this clearly:

> Take an instant of conscious life, suppose memory entirely suppressed, so that the instant yields only the immediate experience it actually contains, you are conceiving the limit of materiality. At the limit there is neither selection nor contraction and consequently no image. Pure perception exists then only in theory, but it performs the kind of practical service which a limiting concept performs in mathematics. There is an infinite approach to it but we can never reach it.[32]

The image explains possible action, but there is no possibility to be considered if action is delimited to the present alone – one action would follow another automatically, without thought, consideration or memory. Deliberation would be impossible in this picture, and so perception would be useless – it would do nothing for action, since action would proceed automatically.

In order to explain this point, it is helpful to consider Bergson's observations on the hierarchy of vertebrates, which develops from basic 'living matter' to more complicated systems of responses to stimuli:

> If we follow, step by step the progress of external perception from the monera to the higher vertebrates, we find that living matter, even as a simple mass of protoplasm, is already irritable and contractile, that it is open to the influence of external stimulation, and answers to it by mechanical, physical and chemical reactions. As we rise in the organic series, we find a division of physiological labor. Nerve cells appear, are diversified, tend to group themselves into a system; at the same time, the animal reacts by more varied movements to external stimulation.[33]

As the nervous system becomes more and more developed, the terminal of stimulus and response becomes more and more complicated, which means that the animal can discern and react to stimuli in more varied manners. What happens when we arrive at human perception above the relatively more automated responses of, say, the jellyfish? Bergson continues the above passage with a blunt illustration of the body's circuit of stimulus and response: the afferent nerve is periph-

The Polarisation of Nature

erally stimulated but 'instead of proceeding directly to the motorcells of the spinal cord and impressing on the muscle a necessary contraction, mounts first to the brain, and then descends again to the very same motor cells of the spinal cord which intervened in the reflex action'.[34] Bergson then asks, what is the nature of this detour? Surely it is not to produce a representation or illustration of movement outside the body. Rather, for Bergson,

> the brain is no more than a kind of central telephonic exchange: its office is to allow communication or to delay it. It adds nothing to what it receives; but, as all the organs of perception send it to their ultimate prolongations, and, as all the motor mechanisms of the spinal cord and of the medulla oblongata have in it their accredited representatives, it really constitutes a center, where the peripheral excitation gets into relation with this or that motor mechanism, chosen and no longer prescribed ... In other words, the brain appears to us to be an instrument of analysis in regard to the movement received and an instrument of selection in regard to the movement executed.[35]

This analysis fits nicely with Bergson's view that perception does not operate with a view to speculation but rather to movement and movement alone. By analysing what it takes in and deciding how it responds, the brain functions as a switchboard operator more than an alchemist of illusions. The brain has no representational powers – it is more akin to the spinal cord than philosophers are wont to admit: 'there is merely a difference of complication, and not a difference in kind, between the functions of the brain and the reflex activity of the medullary system'.[36] Not only the brain, but the entire human body is an image-movement machine oriented toward action instead of knowledge. When discussing perception, Bergson has more in common with contemporary neuroscientists than with philosophers from the eighteenth to twenty-first centuries. The question of correspondence has not been broached because there is no problem of connecting two series of images, one mental and one material, which plagued both idealist and realist philosophers. Bergson's analysis of perception bypasses the question of correspondence in the same way it bypasses the problem of representations – there is only one series of present images, and matter, image and movement are all identified within that series.

The brain's 'telephonic exchange' does, however, install a conscious interval within perception that affords us the possibility of absorbing stimuli and deliberating upon it, in between the reception of an excitation and a response. Unlike inanimate objects, what we receive and how we respond is not simply determined by the automatic procession of material images, but rather is conditioned by the past through memory. As we shall see, the vital difference between organism and matter is duration itself, and the milieu of the organism is composed by duration inserted into material procession. Concrete perception is constituted, then, by the 'telephonic exchange' that selects, orients and regulates the field of images that interests our possible action. As Deleuze writes:

> The brain does not manufacture representations, but only complicates the relationship between a received movement (excitation) and an executed movement (response). Between the two, it establishes an interval (*écart*), whether it divides up the received movement infinitely or prolongs it in a plurality of possible reactions ... By virtue of the cerebral interval, in effect, a being can retain from a material object and the actions issuing from it only those elements that interest him. So that perception is not the object *plus* something, but the object *minus* something, minus everything that does not interest us.[37]

Since the brain and perception do not fabricate representations, there is nothing 'added' to the image; rather, the image is an impoverished actuality, an aspect of actuality selected by our interest. Memory informs this selection process in two distinct ways. The first way (what Bergson calls 'inattentive motor-memory') is through the development of sensory-motor habits, and the second way (what Bergson calls 'attentive image-memory') is through what is often called recollection, or the association of images that can be more or less spontaneously recalled.

A person taking a habitual stroll through their neighbourhood is exercising inattentive motor memory, since his movements proceed unremarkably through the easy recognition of familiar signs. We can be barely conscious of walks or even drives to work because motor-memory is so effectively inattentive. However, if you take a wrong turn and end up in an unfamiliar environment, suddenly your attention shifts and a new kind of perception is required in order to move about the area. The same phenomenon can be witnessed when meeting new people: when we first meet someone we must study them in order to grasp their signs, the way the move their body and the sound of their voice. After several meetings, these signs are all we require in order to recognise them. The point here is that with *inattentive recognition*, perception gives way to effectively automatic responses; whereas with *attentive perception*, we must dwell upon images and absorb them in order to respond to them more effectively.

The Polarisation of Nature

It would be a mistake, however, to think that in recognition we blend perception with images recollected from memory because (and here is an essential point), recollected memory-images can only be recalled *after* recognition, as we can only *recollect* friends and neighbourhoods after we have recognised them. The more accurate statement of the issue is that after recognition, recollected memory-images escort our interaction with friends and neighbourhoods alongside the surprising new signs that demand attention. In this fashion the memories of past encounters informatively tend the image-movements, guiding our selections and responses to them. Bergson writes:

> While external perception provokes on our part movements which retrace its main lines, our memory directs upon the perception received the memory-images which resemble it and which are already sketched out by the movements themselves. Memory thus creates anew the present perception, or rather it doubles this perception by reflecting upon it either its own image or some other memory-image of the same kind.[38]

Memory thus 'reflects' the present image-movement in order to better react to it. The memory-image that is recalled by the perceived image creates a kind of circuit that oscillates between the present image-movements and the past memory-images. When we start a conversation with a friend, for example, we begin by recalling associated memory-images that are recalled by the exigencies of the present image-movements. Our conversation is subsequently guided by the recalled memories, and we move along with our friends in a customary way that is informed by the history of our encounters. Given that history, our attentive recognition, far from being simply a motor-habit, swells with a flow of recollected memories. But if memory-images are not reducible to motor-habits, then they are different from movement-images. The brain may very well be a telephonic exchange but, for Bergson, it is not an electronic catalogue of memory-images. Memory-images are not like perceptual-images at all – they are not merely different in degree, but rather different in kind.

One of Bergson's most controversial theses is that memories are not stored within the brain. But while controversial, this thesis is not surprising: since memory-images differ in kind from perception-images, they cannot be understood in terms of movement. Thus, there is a kind of dualism in images, albeit one drawn within time itself: perception treats image-movements that are present by definition, since the present is material, while memory treats recollected-images that are past, and so not extensive or material movements. This distinction is important to uphold, because if memory-images could blend with perceptual images, then perception would no longer be tied to the plane of image-movements, and we would once again relapse into the representational account of perception – perceptions would be mixed with memories inside the brain, reasserting the prepositional muddle that divides images into two series that merely differ in degree, a mental series of memory-images inside the brain that combine with external movements outside the brain in order to produce the phenomena of perception. Such a blending of memory-images and perceptual-images would simply amount to another version of mental images or representations, once again separating the mind from matter in terms of space. This would reassert the bifurcation and confuse both perception and memory alike. Thus it is crucial for Bergson to maintain a difference in kind between memory-images and image-movements, which entails that memory-images cannot be located inside the brain or within any other image-movement. Instead of drawing a distinction between the mind and matter in terms of space, Bergson will distinguish memory from matter in terms of time.

The Metaphysics of Time and Matter

Sensory-motor habits, for Bergson, reveal a 'tendency of every organism to extract from a given situation that in it which is useful ... that it may serve situations of the same kind'.[39] And while the recollection of distant memories may seem to have little utility in the present moment, by and large the function of memory is crucially oriented toward recognition of present scenarios and the management thereof. Neuroscientist Robert Stickgold describes an experiment where even the wild reveries of sleep come to perform a service for the present. Participants in an experiment were placed in a video-game machine that simulated the experience of skiing. Their performances were recorded and evaluated, and they remained at a test centre overnight so that their sleep could be monitored. At certain intervals in their REM cycles, the participants were awoken and asked if they could recall what they had been dreaming about. Their dreams were often relevant to the day's events – 'I was walking through snow', one participant reported – and the participants who continued to practice on the machine without having slept had significantly poorer performances than those who had slept and dreamt before getting back on the simulator.[40] Stickgold muses that it was as if the mind were saying, 'here's what I know about *walking* through snow, how can this help me *ski* through snow?' Memories and dream images do not fuse with perceptual images but rather direct our selection of important perceptual images, allowing us to better conduct our eventual action upon our environment.

The memorial past, for Bergson, is a depth inserted into the plane of present images that allows perception to distinguish the marks which it finds relevant. The selection of images from the theoretical infinity of possible percepts, then, is a result of habits developed in the past and material movements available for capture. One early commentator describes the selection of a colour image as follows:

> Take the case of red colour sensation. According to the physical theory the condition of this sensation is the propagation of 451 billion vibrations per second and the smallest part of a second in which I can distinguish sensations is the 1/500th. Let us accept the figures. What then? Colour sensation depends physically on the number of these vibrations contracted into our psychical moment of duration. Vary the number and we vary the colour. Precisely, says Bergson, but is there not also another way in which we could attain the same result? Suppose we relax the psychical tension. Suppose that without the physical rhythm of the propagation altering at all our grasp on it is relaxed, so that the tension extends. Must we not suppose that the effect would be precisely the same? The colour would change and finally disappear with the approach to the limit of coincidence with pure vibrations.[41]

This example reveals that the perception of colour is reducible to neither the physical nor the habitual components at work in its selection but, rather, that perception of colour is constituted by the calibration of both the physical and psychical components that result in perception. Again, we do not 'add' anything to material images, but rather focus on certain elements in order to form a picture that interests our eventual action: '*there is in matter something more than, but not something different from, that which is actually given*'.[42] Deleuze reiterates this point when writing that 'perception is not the object *plus* something, but the object *minus* something, minus everything that does not interest us'.[43] So the object of perception is a remainder of an elaborate process of memory, habit and motor-sensations that capture only a choice sliver of movements that an organism deems relevant for its own qualitative adventure. Bergson sometimes discusses this process in terms of a transition from the *virtual* to the *actual*:

> To transform [something's] existence into representation, it would be enough to suppress what follows it, what precedes it, and also all that fills it, and to retain only its external crust, its superficial skin. That which distinguishes it as a *present* image, as an objective reality, from a *represented* image is the necessity which obliges it to act through every one of its points upon all the points of all other images, to transmit the whole of what it receives, to oppose to every action an equal and contrary reaction, to be, in short, merely a road by which pass in every direction, the modifications propagated throughout the immensity of the universe. I should convert it into representation if I could isolate it, especially if I could isolate its shell. Representation is there, but always virtual – being neutralised, at the very moment when it might become actual, by the obligation to continue itself and to lose itself in something else. To obtain this conversion from the virtual to the actual, it would be necessary, not to throw more light on the object, but, on the contrary . . . to diminish it by the greater part of itself, that the remainder, instead of being encased in its surroundings as a *thing*, should detach itself from them as a *picture*.[44]

A *thing* is the present image plus all of its material connections, its 'modifications propagated throughout the immensity of the universe'. A picture, on the other hand, is *that same image* minus its complicated actual involvement in the universe, as it concerns only our own possible action. Bergson reiterates this point in *Creative Evolution*:

> The distinct outlines which we see in an object, and which give it its individuality, are only the design of a certain kind of influence that we might exert on a certain point of space: it is the plan of our eventual actions that is sent back to our eyes, as though by a mirror, when we see the surfaces and edges of things. Suppress this action, and with it consequently those main directions which by perception are [sic] traced out for it in the entanglement of the real, and the individuality of the body is re-absorbed in the universal interaction which, without doubt, is reality itself.[45]

Perception is a kind of capture or cut in the continuation of image-movements, the detachment of a picture from the immense material involvement of a thing's physicality. In another passage, Bergson describes perception as a reflection of luminosity effected by the interaction of two media of different densities – the perceptual apparatus and the material movements:

> When a ray of light passes from one medium into another, it usually traverses it with a change of direction. But the respective densities of the two

media may be such that, for a given angle of incidence, refraction is no longer possible. Then we have total reflection. The luminous point gives rise to a *virtual* image which symbolises, so to speak, the fact that the luminous rays cannot pursue their way. Perception is just a phenomenon of the same kind.[46]

The image, then, is precisely where it appears to be – not inside the mind, but at the calibrated point of interaction between two media that produces a virtual image, held in place by the effort (or density, if you prefer) of memorial habit that selects and the density of material movements available for selection. In this fashion, Bergson can claim that a single series of material movement-images operates upon two different metaphysical registers – the past of memory and consciousness which is not material and the present of material images. It makes no sense, then, to inquire as to whether perceptual images are inside or outside of us. 'Every image is within certain images and without others; but of the aggregate of images we cannot say that it is within us or without us, since interiority and exteriority are only relations among images.'[47] Bergson is fond of saying that containers do not issue from content.[48] Instead of postulating non-extensive images in the mind and extensive images in space, Bergson discusses the image as a calibrated result of myriad contractions, involving both the memorial influence of habit and the physical movements of matter.

But what, then, is the real distinction between mind and matter for Bergson? After all, by asserting that memory-images are different in kind from image-movements as the mind is different from matter, does Bergson not inadvertently fall back into a bifurcation of nature? There is certainly a dualism in Bergson, but it is a dualism of time and not space. Matter is extensive, but the plane of extension is understood as the present – itself a dimension of time. Memory and consciousness are certainly not extensive, but rather intensive depths staked in the past. The difference between mind and matter is thus a difference between the past and the present. Nature is not so much bifurcated, then, as much as it is polarised upon a continuum of time, where the present is a peak of duration at its most contracted state, and the past is the expanded or relaxed depth of duration. Bergson offers what is perhaps the clearest formulation of the distinction between the past and the present in terms of the distinction between the subject and an object.

The Distinction Between Subject and Object

'Our perception of an object distinct from our body, separated from our body by an interval', Bergson writes, 'never expresses anything but a *virtual* action.'[49] But the closer objects come, the more pressing their action upon us becomes, and the more virtual action becomes real action. The difference between the subject and object, then, is a difference in *time* – it is the difference between possible or virtual action and actual movement. Bergson provocatively claims that 'questions relating to subject and object, to their distinction and their union, should be put in terms of time rather than of space'.[50] The difference between the subject and an object is that a subject can absorb time, postponing action. Consciousness itself is the memorial consideration of events and affects – the qualitative experience of consciousness is an affect of time, occurring within intervals of action. Conscious sensations, then, are the consequences of absorbed action, and they express the fact that time is endured by the living body in a way that it cannot be endured by a non-living body. The living body, Bergson writes:

> is not a mathematical point; it is a body, exposed, like all natural bodies, to the action of external causes which threaten to disintegrate it ... it does not merely reflect action received from without; it struggles, and thus absorbs some part of this action. Here is the source of affection. We might therefore say, metaphorically, that while perception measures the reflecting power of the body, affection measures its power to absorb.[51]

While perception reflects the virtual action of the body upon its environment, affection is the action (actually) suffered by the body from its environment. Bergson reiterates this distinction between perception and sensations or affections in terms of virtual and real action: 'Our sensations are, then, to our perceptions that which the real action of our body is to its possible, or virtual, action. Its virtual action concerns other objects and is manifested within those objects; its real action concerns itself, and is manifested within its own substance.'[52] Perception is the function of the organism that arrests virtual images from the series of image-movements according to the contours of the organism's interest. The cost of this interest, however, is the felt investment of the organism's centrality and interiority. We perceive the rain before we get wet, and the virtual image of rain is possible action, the affect of being wet is the actual action. The surface of the body thus marks the 'common limit of the external and the internal . . . the only portion of space which is both perceived and felt'.[53] But the distinction between the internal and the external, then, is not so much a question of space but rather a question of time and affection – it is the product of a qualitative difference between the endured action from a body's environment and the possible action of a body upon its environment. The limit between the inside and the outside of a living body is a primarily temporal limit between the durational past and the actual present.

Behind the distinction between the subject and the object, then, lies a distinction within time itself. For Bergson, objects can differ in degree, like one present moment can differ from another. But image-movements are by definition tied to the present, and movement can only differ from movement in terms of degree. The present thus forms a kind of clock time – time understood in terms of movement. But the past is not understood in terms of movement, but rather intensity. The difference between intensive duration and extensive movement is thus a difference in kind – differences in degree obtain only within the homogeneous time of image-movements. In *Time and Free Will*, Bergson discusses the crucial distinction as follows:

> When I follow with my eyes on the dial of a clock the movement of the hand which corresponds to the oscillations of the pendulum, I do not measure duration, as seems to be thought; I merely count simultaneities, which is very different. Outside of me, in space, there is never more than a single position of the hand and the pendulum, for nothing is left of the past positions. Within myself a process of organization or interpenetration of conscious states is going on, which constitutes true duration. It is because I endure in this way that I picture to myself what I call the past oscillations of the pendulum at the same time as I perceive the present oscillation.[54]

Bergson does not explicitly discuss technology often, so the claim that the dial of an analogue clock trains our understanding of time is already remarkable. Behind it we can decipher the thesis that technological artifices (like the clock) inform our concepts of natural qualities (like time). In particular, the claim is that our idea of succession is traced on the outlines of the movement of a clock, occluding the conscious duration that renders succession possible. That is, succession itself is a product of consciousness, since for material objects there is only one position, and hence no succession. It is only by retaining the past, by absorbing action, in short, by enduring, that succession can exist. *In pure space there is only juxtaposition and simultaneity.* Space is only in the present, and consciousness only in the past. The problem is that we tend to conflate the qualitative states of consciousness and the spatial movements of the clock, reading the intensive process of duration (and the cogito) through the extensive movements of space (and the clock). Bergson writes:

> As the successive phases of our conscious life, although interpenetrating, correspond individually to an oscillation of the pendulum which occurs at the same time, and as, moreover, these oscillations are sharply distinguished from one another, we get into the habit of setting up the same distinction between the successive moments of our conscious life: the oscillations of the pendulum break it up, so to speak, into parts external to one another: hence the mistaken idea of a homogeneous inner duration, similar to space, the moments of which are identical and follow, without penetrating, one another.[55]

We thus conceive of internal duration, the time of consciousness, as a spatial and homogeneous time. This transforms the duration of consciousness into the space of the pendulum – time becomes simply the measure of movement, and the qualitative difference in kind between the past and the present is forgotten. A merely symbolic representation of time is thus taken for time itself, and space becomes the measure of time, when in fact, for Bergson, it is time that conditions space.

The difference between the Cartesian bifurcation of nature and the Bergsonian polarisation of nature hinges upon the distinction between two different forms of time. For Descartes, time was understood in terms of space – time was homogeneous, clock time. For Bergson, however, clock time is simply movement, and real duration is a qualitative and heterogeneous multiplicity. It may seem strange to talk about a 'multiplicity' as heterogeneous or qualitative, since the idea of a multiple suggests a numerical addition of homogeneous units. Yet Bergson intends to indicate two very different kinds of multiplicities – one the one hand, a numerical and homogeneous multiplicity and, on the other, a qualitative and a heterogeneous multiplicity – the former is countable, a quantity traced upon the outlines of extensive space, and the latter is innumerable, a qualitative series of changes that interpenetrate and transform each other. Bergson writes:

> We should therefore distinguish two forms of multiplicity, two very different ways of regarding duration, two aspects of conscious life. Below homogeneous duration, which is the extensive symbol of true duration, a close psychological analysis distinguishes a duration whose heterogene-

ous moments permeate one another; below the numerical multiplicity of conscious states, a qualitative multiplicity; below the self with well-defined states, a self in which *succeeding each other* means *melting into one another* and forming an organic whole. But we are generally content with the first, i.e. with the shadow of the self projected into homogeneous space.[56]

Consciousness forms a whole, yet it continually separates itself artificially by tracing itself on the outlines of movement, on the outlines of clock time. We habitually think of consciously endured moments as if they were successive instants ('homogeneous duration'), but it is rather duration ('true duration') that is the basis of succession. Bergson often refers to the organic whole of consciousness as the fundamental self, which subtends the artificial impression of life as a series of distinct instants. To be sure, there are diverse moments but, upon closer analysis, Bergson finds that they are not reducible to a numerical succession of instants: the composition of each moment reveals a depth of duration that is itself the principle of each moment's eventuality.

The conflation of these two multiplicities has caused both philosophy and science considerable misunderstandings. It produces the bifurcation with all its false problems, most notably the impasse between realism and idealism. The illusion, however, is not so easy to dispel – it is based, as Deleuze writes, 'in the deepest part of the intelligence', which 'is not, strictly speaking, dispelled or dispellable', but rather 'can only be *repressed*'.[57] It is only with considerable difficulty that we think of consciousness as intensive duration, when outwardly time seems to merely be the succession of movements in space. But Bergson believed in the possibility of a new concept of nature that both science and philosophy could recognise: 'science and consciousness are agreed at bottom, provided that we regard consciousness in its most immediate data and science in its remotest aspirations'.[58] This new concept of nature would be a marriage between the method of intuition and the abstractions of scientific measurement. In fact, Bergson's project is to supplement the abstractions of science with the method of intuition. The ultimate goal of Bergson's metaphysics is to connect matter and memory within the register of life and organism, and thus to reclaim the 'natural articulations' obfuscated by the artifice of representation.[59] By 'natural articulations' Bergson means the real or 'natural' world order of events that compose both experience and knowledge, both scientific and philosophical. In this sense, Bergson's metaphysics, with its distinction between differences of degree and differences in kind, matter and memory, apprehends natural articulations and thereby reveals the process that *artificially produces* the notion of independently represented bodies in discontinuous space. This, in turn, encourages Bergson to hope for a reconciliation between metaphysics and science. He writes:

> A *moving continuity* is given to us, in which everything changes and yet remains: why then do we dissociate the two terms, permanence and change, and then represent permanence by *bodies* and change by *homogeneous movements* in space? This is no teaching of immediate intuition; but *neither is it a demand of science* [italics mine]. The object of science is, on the contrary, to rediscover the natural articulations of a universe we have carved artificially.[60]

It is clear that for Bergson science and intuition are ultimately reconciled by their orientation toward natural articulations, and the artificial representation of nature is an obfuscation that threatens philosophy and science mutually. Deleuze's appendix to *Bergsonism* emphasises this complementarity between intuition and science: 'Bergson did not merely criticise science as if it went no further than space, the solid, the immobile . . . science is never "reductionist" but, on the contrary, demands a metaphysics – without which it would remain abstract, deprived of meaning or intuition.'[61]

The Eleatic Mathematician: Zeno's Paradox

Bergson's distinction between two multiplicities (which, as we shall see in the following chapter, Deleuze makes pivotal use of when discussing difference and representation), is derived from G. B. R. Riemann's distinction between discrete and continuous multiplicities. *Discrete multiplicities* are homogeneous and numerical multiplicities whose items display the principle of their own metric *a priori*. For example, with a ruler whose metric is the inch, one may measure any of the 'multiplicity' of numerical values stretching along the length of the ruler in terms of the inch, precisely because it contains its own metric *a priori*. Similarly, one may convert one unit of American currency, say a dollar, into other units of the same currency (ten dimes, for instance), because a homogeneous numerical measure persists throughout. *Continuous multiplicities*, on the other hand, are heterogeneous and non-numerical multiplicities where the metric itself changes alongside its divisions. In this case, the ground of the metric arrives from outside of the multiplicity, from the diverse forces acting within it – for example, while the dollar holds its numerical metric in America, the international exchange rate fluctuates according to non-numerical and heterogeneous forces of value.[62] Riemann writes:

> in a discrete manifold, the principle or character of its metric relations is already given in the notion of the manifold, whereas in a continuous manifold this ground has to be found elsewhere, i.e. has to come from outside. Either, therefore, the reality which underlies space must form a discrete manifold, or we must seek the ground of its metric relations (measure-conditions) outside it, in binding forces which act upon it.[63]

When we represent nature extensively, we deploy something of a global metric – something of a Cartesian coordinate space. However, when we approach nature intuitively – as Bergson does – then we find a depth of duration that is beyond such metrics, something that can only be felt and not measured. Thus, the heterogeneous multiplicity of conscious duration lies beyond the capture of representation. Yet we are constantly representing concrete duration (continuous multiplicity) in terms of abstract metrics (discrete multiplicities), thereby producing a number of false problems that prevent both science and philosophy from developing an adequate concept of nature. Perhaps the most famous of such false problems is Zeno's Achilles paradox. A brief examination of this paradox sheds considerable light on what is at stake in Bergson's distinction between multiplicities.

The second of Zeno of Elea's paradoxes on motion features Achilles, the celebrated warrior, losing a race to a tortoise. None of Zeno's original writings are extant, but Simplicius recounts the paradox in his commentary on Aristotle's *Physics*, one among many ancient philosophers who attempted a solution to the paradoxes:

> The [second] argument was called 'Achilles', accordingly, from the fact that Achilles was taken [as a character] in it, and the argument says that it is impossible for him to overtake the tortoise when pursuing it. For in fact it is necessary that what is to overtake [something], before overtaking [it], first reach the limit from which what is fleeing set forth. In [the time in] which what is pursuing arrives at this, what is fleeing will advance a certain interval, even if it is less than that which what is pursuing advanced ... And in the time again in which what is pursuing will traverse this [interval] which what is fleeing advanced, in this time again what is fleeing will traverse some amount ... And thus in every time in which what is pursuing will traverse the [interval] which what is fleeing,

being slower, has already advanced, what is fleeing will also advance some amount. (Simplicius(b) *On Aristotle's Physics*, 1014.10)

Of course, the history of philosophy offers responses this paradox. What is interesting about Bergson's response is that instead of attempting to 'solve' the paradox, he shows how the paradox is produced. In other words, Bergson reveals the underlying process subtending the formulation of the paradox.

The paradox operates by playing the spatial limits of the tortoise's advance against the temporal interval of Achilles' pursuit. If we take the tortoise's trajectory as a line AB, then Achilles must advance to B in order to overtake the tortoise. A small feat, but during the interval of Achilles' advance, the tortoise has shockingly extended the terminal point, albeit slightly, to AC, setting a new limit for Achilles' required progress. Again, by the time Achilles reaches point C, a new limit, AD, has been established, and so on. The paradox, then, takes the path that must be traversed by the trailing Achilles and identifies it with Achilles' (and the tortoise's) movement. The fallacious assumption, for Bergson, is that Achilles' and the tortoise's

> movement coincides with their path and we may divide it, like the path itself, in any way we please. Then, instead of recognizing that the tortoise has the pace of a tortoise and Achilles the pace of Achilles, so that after a certain number of these indivisible acts or bounds Achilles will have outrun the tortoise, the contention is that we may disarticulate as we will the movement of Achilles and, as we will also, the movement of the tortoise: thus reconstructing both in an arbitrary way, according to a law of our own which may be incompatible with the real conditions of mobility.[64]

In other words, we disarticulate movement from its natural qualitative constitution by reconstructing it in a homogeneous spatial grid, which introduces the commonality by which Achilles can be measured alongside the tortoise. Each leap of Achilles is concretely an indivisible event – Achilles does not move first an inch, and then another inch, when advancing upon the tortoise. In fact, the path of Achilles is not even remotely similar to the path of the tortoise – the patch of ground that Achilles selects to step on, the heat that he manages from the blinding sun, and all the other environmental factors that are virtually arrested by his constitution are a far cry from the environmental design of the tortoise, who composes a different path from its immense thirst amid the texture and dryness of the earth. Yet the Eleatic mathematician reconstructs their respective movements in terms of a homogeneous and quantitative length. Bergson calls this reconstruction arbitrary because the divisions used are arbitrary – instead of a race composed of the concrete leaps of Achilles and the concrete waddles of the tortoise, both with their respective velocities, environmental evaluations and rates of change, we reconstruct the race in terms of homogeneous and arbitrary spatial units (feet or inches) alongside homogeneous and arbitrary temporal units (minutes or seconds). We thus artificially reduce a qualitative multiplicity into a quantitative one.

William James, in his lectures gathered into *A Pluralistic Universe*, discusses the paradox as it concerns Bergson in terms of fractions: 'Expressed in bare numbers, it is like the convergent series ½ plus ¼ plus ⅛. . ., of which the limit is one. But this limit, simply because it is a limit, stands outside the series, the value of which approaches it indefinitely but never touches it.'[65] Fractions give us an infinite divisibility between wholes or units. Configuring movement in terms of such infinite divisibility, however, artificially divides what concretely is a qualitatively indivisible leap of Achilles. James is thus able to playfully conclude, in his characteristic manner: 'but in point of fact nature doesn't make eggs by making first half an egg, then a quarter, then an eighth, etc. and adding them together. She either makes a whole egg at once or none at all.'[66] Similarly, Achilles makes a single leap or none at all – anything else is an abstraction from concrete reality.

In the *Critique of Pure Reason*, Kant acknowledged disarticulation as a condition for the mathematical construction of the concepts of space and time: 'space and time, and a concept of these, as quanta, can be exhibited *a priori* in pure intuition, i.e., constructed, together with either its quality (its shape) or else merely its quantity (the mere synthesis of the homogeneous manifold) through number'.[67] This disarticulation actually ensured that 'mathematics traveled the secure path of a science' because geometers realised that in analysing the properties of figure we should not start with a concrete figure but rather *produce* the figure from what it definitively required.[68]

It is in this sense that 'mathematical definitions can never err' because 'the concept is first given through the definition, it contains just that which the definition would think through it'.[69] Kant frequently describes mathematical concepts as arbitrary precisely because of the disarticulation that allows for their construction. Lisa Shabel illustrates this point clearly:

> For example, to attempt to define the concept triangle one considers the possibility of constructing a three-sided rectilinear figure. Kant thinks of this concept as 'arbitrary' in the following sense: in considering such a concept, one knows precisely what its content is since one 'deliberately made it up', and, moreover, the concept was not 'given through the nature of the understanding or through experience' (A 729/B 757) ... in the case of a triangle, one considers the concept figure (that which is contained by any boundary or boundaries) together with the concepts straight line and three, and then proceeds to effect the synthesis of these concepts by exhibiting an object corresponding to this new concept.[70]

In this fashion we might arbitrarily create any shape whatsoever (we can determine the angles and properties of even a Cartesian chiliagon) because there is nothing in the concept that results from experience. Thus, mathematical concepts are definable, they are 'closed' concepts delimited by the arbitrary choosing of factors to feature within the concept. But this disarticulation also facilitates the commensurability between Achilles and the tortoise, in the sense that they can be compared to each other through the homogenising measure of spatial length. From the homogenisation of this measurement arises the same tendency to abstract from lived movement and reconstitute it in the figure of a coordinate grid taken at an instant, where each part can be mathematically brought to bear on each other part.[71] There are no differences in kind on a coordinate grid, all differences are simply differences of degree.

This is precisely the conceptual framework that obtains when we consider space traversed in terms of a geometrical arrangement of points, lines and figures that changes from one instant to another. We are used to hearing the phrase, 'at time t', in discourses subscribing to this geometrical model of space. The problem, Bergson writes, is that 'when the mathematician calculates the future state of a system at the end of a time t, there is nothing to prevent him from supposing that the universe vanishes from this moment till that, and suddenly reappears'.[72] The mistake lies in the conception of space as an abstract grid and time as a series of non-enduring instants through which that grid somehow passes, reducing movement within the grid to so many incremental shifts.[73] Movement, in other words, is misconstrued when understood in terms of abstract and homogeneous time.[74] What is lost in the spatial understanding of time, as exemplified in Zeno's paradox, is change itself, which happens, as it were, in between the two limit-instants. Achilles overtakes the tortoise because his body moves in a register of duration that cannot be accounted for by way of non-enduring instants. The Eleatic view, the geometer-mathematician's view, has a certain precision when analysing spatial configurations upon a coordinate grid, but the numerical reductions of movement are discontinuous analyses of movement, and 'you cannot make continuous being out of discontinuities'.[75] Hence the requirement to distinguish between, on the one hand, a spatial, homogeneous and quantitative multiplicity, and on the other hand, a durational, heterogeneous and qualitative multiplicity:

> [Spatial multiplicity] is a multiplicity of exteriority, of simultaneity, of juxtaposition, of order, of quantitative differentiation, of *difference in degree*; it is a numerical multiplicity, *discontinuous and actual*. The other type of multiplicity appears in pure duration: It is an internal multiplicity of succession, of fusion, of organization, of heterogeneity, of qualitative discrimination, or of difference in kind; it is a virtual and continuous multiplicity that cannot be reduced to numbers.[76]

For Bergson, the philosophical method of intuition reveals the qualitative differences that naturally articulate movement, while the scientific methods of measurement reveal quantitative differences that artificially articulate movement. Both are required in order to understand nature, yet there is a difference in kind between them that, if unacknowledged, leads to false problems. The qualitative difference of duration transcends the quantifying intellect, yet it nonetheless subtends the intellectual apprehension of time. The problem lies in forgetting that underneath the discrete quantification of time is the irreducibly qualitative continuity of duration. As Bergson writes, '[Intellect] does not like what is fluid, and solidifies everything it touches. We do not *think* real time. But we *live* it, because life transcends intellect.'[77]

Notes

1. Martin Heidegger, *Basic Writings*, ed. David Farrell Krell (New York: HarperCollins Publishers, Inc., 1993), p. 292.
2. German Idealism and Naturphilosophie are particularly salient examples in this regard. See Jeremy Dunham, Iain Hamilton Grant and Sean Watson, *Idealism: The History of a Philosophy* (Montreal: McGill-Queen's University Press, 2011). For the comprehensive and magisterial history admired by Nietzsche, see Frederick Lange, *The History of Materialism, and Criticism of its Present Importance (in three volumes)*, 2nd edn (New York: Routledge, 2000).

3. Gilles Deleuze, *Cinema 1: The Movement-Image*, trans. H. Tomlinson and B. Habberjam (Minneapolis: University of Minnesota Press, 1986), p. 56. The same claim was also made two years earlier in Deleuze's Cours Vincennes – St Denis: Bergson, *Matière et Mémoire* – 05/01/1981.
4. Bergson, *Matter and Memory*, p. 10.
5. Bergson, *Matter and Memory*, p. 11.
6. Bergson, *Matter and Memory*, p. 17.
7. Bergson, *Matter and Memory*, p. 19.
8. Bergson, *Matter and Memory*, p. 9.
9. On this point, and for a critical analysis of Bergson in general, see Anthony Edward Pilkington, *Bergson and His Influence: A Reassessment* (Cambridge: Cambridge University Press, 1976).
10. Bergson, *Matter and Memory*, pp. 19–20.
11. Bergson, *Matter and Memory*, p. 20.
12. Bergson, *Matter and Memory*, p. 22.
13. Bergson, *Matter and Memory*, pp. 21–2.
14. Bergson, *Matter and Memory*, p. 23.
15. For a phenomenological interpretation of the role of the body in perception, see Leonard Lawlor, *The Challenge of Bergsonism: Phenomenology, Ontology, Ethics* (New York: Continuum, 2003).
16. Bergson, *Matter and Memory*, pp. 23–4.
17. Bergson, *Matter and Memory*, p. 12. Parentheses mine.
18. Bergson, *Matter and Memory*, p. 25.
19. Bergson, *Matter and Memory*, p. 28.
20. Bergson, *Matter and Memory*, p. 45.
21. Bergson, *Matter and Memory*, p. 9.
22. Deleuze, *Cinema 1*, p. 58.
23. Deleuze, *Cinema 1*, p. 58.
24. Oliver Sacks, *An Anthropologist on Mars: Seven Paradoxical Tales* (New York: Random House, Inc., 1995), p. 115.
25. Sacks, *An Anthropologist on Mars*, p. 110.
26. Sacks, *An Anthropologist on Mars*, p. 127.
27. Sacks, *An Anthropologist on Mars*, p. 133.
28. Sacks, *An Anthropologist on Mars*, p. 137.
29. Sacks, *An Anthropologist on Mars*, p. 138.
30. Bergson, *Matter and Memory*, p. 33.
31. There is, in fact, no 'pure perception' of Peter, as recognition is a function of memory which informs perception. Bergson cites several cases of aphasia where recognition is dysfunctional as support for this claim.
32. H. Wildon Carr, 'What does Bergson Mean by Pure Perception?', *Mind*, New Series, Vol. 27, No. 108 (October 1918), pp. 472–4.
33. Bergson, *Matter and Memory*, pp. 28–9.

34. Bergson, *Matter and Memory*, p. 29.
35. Bergson, *Matter and Memory*, p. 30.
36. Bergson, *Matter and Memory*, p. 29.
37. Gilles Deleuze, *Bergsonism*, trans. Hugh Tomlinson and Barbara Habberjam (New York: Zone Books, 1988), pp. 24–5.
38. Bergson, *Matter and Memory*, p. 101.
39. Bergson, *Matter and Memory*, p. 167.
40. J. M. Ellenbogen, J. D. Payne, Robert Stickgold, 'The Role of Sleep in Declarative Memory Consolidation: passive, permissive, active or none?' *Curr Opin Neurobiology*, Vol. 16, No. 6 (December 2006), pp. 716–22. Epub 7 Nov. 2006.
41. Carr, 'What does Bergson Mean by Pure Perception?', p. 474.
42. Bergson, *Matter and Memory*, p. 71.
43. Deleuze, *Bergsonism*, p. 25.
44. Bergson, *Matter and Memory*, pp. 35–6. The reader will note that Bergson distinguishes 'representation' into 'virtual' and 'actual' representations. We move from virtual representations to actual representations through a process of diminishing, a process of 'picturing' a 'thing'. As I am using Bergson to argue against what I have called 'representation', I will refer to 'virtual representations' as 'things' and 'actual representations' as 'pictures', with notes where context or further distinction is required.
45. Henri Bergson, *Creative Evolution*, trans. A. Mitchell (Mineola: Dover Publications, 1998), p. 11.
46. Bergson, *Matter and Memory*, p. 37.
47. Bergson, *Matter and Memory*, p. 25.
48. Bergson, *Matter and Memory*, p. 41.
49. Bergson, *Matter and Memory*, p. 57.
50. Bergson, *Matter and Memory*, p. 71; italics mine.
51. Bergson, *Matter and Memory*, p. 56.
52. Bergson, *Matter and Memory*, p. 57.
53. Bergson, *Matter and Memory*, p. 57.
54. Henri Bergson, *Time and Free Will: An Essay on the Immediate Data of Consciousness*, trans. F. L. Pogson (New York: Dover Publications, Inc., 2001), p. 108.
55. Bergson, *Time and Free Will*, p. 109.
56. Bergson, *Time and Free Will*, p. 128.
57. Deleuze, *Bergsonism*, p. 21.
58. Bergson, *Matter and Memory*, p. 197.
59. Bergson, *Matter and Memory*, p. 197.
60. Bergson, *Matter and Memory*, p. 197.
61. Deleuze, *Bergsonism*, p. 116.
62. Weyl states Riemann's position as follows: 'space in itself is nothing more than a three-dimensional manifold devoid of all form; it acquires

a definite form only through the advent of the material content filling it and determining its metric relations'. Hermann Weyl, *Space–Time–Matter*, trans. H. Brose (New York: Dover Publications, Inc., 1922), p. 98. Also see Arkady Plotnitsky, 'Algebras, Geometries and Topologies of the Fold: Deleuze, Derrida and Quasi-Mathematical Thinking', in Paul Patton and John Protevi (eds), *Between Deleuze and Derrida* (London: Continuum, 2003), pp. 98–119 (especially pp. 101–3).

63. Weyl provides Riemann's own words regarding the distinction between discrete and continuous manifolds in *Space-Time-Matter*, p. 97. While the preposition 'outside' here unfortunately suggests a spatial location, we understand it as indicating a 'beyond' with regard to the discrete multiplicity, and so Deleuze is able to reconfigure it in terms of intensity, a multiplicity 'within' that is nonetheless 'beyond' the extensive.
64. Bergson, *Matter and Memory*, p. 192.
65. William James, *A Pluralistic Universe* (New York: Longmans, Green, and Co., 1909), p. 230.
66. James, *A Pluralistic Universe*, p. 230.
67. Kant, *Critique of Pure Reason*, A 720/B 748.
68. Kant, *Critique of Pure Reason*, Preface to second edition, B xi–xii.
69. Kant, *Critique of Pure Reason*, A 731/B 759.
70. Lisa Shabel, 'Kant's Philosophy of Mathematics', in *The Cambridge Companion to Kant and Modern Philosophy*, ed. P. Guyer (Cambridge: Cambridge University Press, 2006).
71. Deleuze comments on this in *Difference and Repetition*: 'experimentation constitutes relatively closed environments in which phenomena are defined in terms of a small number of chosen factors (a minimum of two – for example, Space and Time for the movement of bodies in a vacuum). Consequently, there is no reason to question the application of mathematics to physics: physics is already mathematical, since the closed environments or chosen factors also constitute systems of geometrical co-ordinates. In these conditions, phenomena necessarily appear as equal to a certain quantitative relation between the chosen factors. Experimentation is thus a matter of substituting one order of generality for another' (p. 3).
72. Bergson, *Creative Evolution*, p. 22.
73. Whitehead makes a similar point in *Principles of Natural Knowledge*: 'The ultimate fact embracing all nature is (in this traditional point of view) a distribution of material throughout all space at a durationless instant of time, and another such ultimate fact will be another distribution of the same material throughout the same space at another durationless instant of time . . . Some modification is evidently necessary. No room has been left for velocity, acceleration, momentum, and kinetic energy, which certainly are essential physical quantities.'

A. N. Whitehead, *An Enquiry Concerning the Principles of Natural Knowledge* (Cambridge: Cambridge University Press, 1925), p. 2.
74. There is a perhaps apocryphal, but nonetheless exemplary, tale sometimes told about the antagonism between Bishop Berkeley and Newton that sheds some light on how different mathematical systems reconstruct movement. Berkeley was curious as to how Newton was calculating planetary orbits with such success and precision, as the line equations and slope forms for measuring change with the algebra of their day required an incredible amount of data in order to produce results comparable to Newton's. Measuring the planetary orbits with those linear tools would amount to a tedious exercise in approximation, if only because one would be calculating curved trajectories through a serious of measured slopes of straight lines, statically comprehended. Newton, with his famed secrecy, had kept his newfound calculus hidden from his colleagues, while nonetheless employing its predictive powers for rates of change in orbital curves. When pressed to deliver his procedure, Newton apparently tried to translate his calculus into the familiar algebraic equations, but certain problems became manifest in the translation. If algebra works through static comprehension, calculus operates dynamically, and so his calculations, arising as they did from derivatives, limits and rates of change, lost the dynamism that was better suited for calculating movement when reconfigured in the static terms of algebra. His results, then, remained suspicious, if not miraculous, until Leibniz published his calculus, forcing Newton to follow suit. To be sure, the distinction between calculus and algebra, or 'transcendental' and 'algebraic' functions, is not as clear-cut as it is depicted in this tale of Berkeley and Newton. In fact, much of modern algebraic geometry is an effort to give algebraic meaning to transcendental notions without resorting to transcendental methods. I mention this tale simply because it indicates two very different ways of reconstructing movement mathematically. Daniel Smith has an excellent essay on the different trajectories of static mathematical multiplicities, such as those we find in set theory, and the dynamic multiplicities of calculus. By examining these different multiplicities, Smith illuminates a crucial difference between Badiou and Deleuze in terms of their mathematical persuasions. See Daniel Smith, 'Mathematics and the Theory of Multiplicities: Badiou and Deleuze Revisited', *Southern Journal of Philosophy*, Vol. 41, No. 3 (Fall 2003), pp. 411–49.
75. James, *A Pluralistic Universe*, p. 236.
76. Deleuze, *Bergsonism*, p. 38.
77. Bergson, *Creative Evolution*, p. 46.

PART II
Toward a New Philosophy of Nature

3

Difference and Representation: Deleuze and the Reversal of Platonism

The distinction between two multiplicities in the previous chapter highlighted the two different ways that we can approach nature. Broadly speaking, we can approach nature as if it were primarily a spatial entity, which leads to bifurcation, or we can approach it as a primarily temporal process, as is the case with polarisation in Bergson. Deleuze shares Bergson's critical evaluation of the bifurcation of nature as well as his emphasis on time and duration. As we shall see, Deleuze's concept of difference, which is a crucial component of his philosophy of time and nature, can be profitably examined within the lineage of both Whitehead and Bergson, precisely because all three endeavour to construct a philosophy of nature based on time. The problem, Deleuze finds, is that philosophy since Plato has been constrained by thinking of entities like time, nature, the subject and the object in terms of identity. This chapter examines how Deleuze, through a reversal of Platonism, formulates a concept of difference apart from identity, and how difference apart from identity relates to a notion of time and becoming beyond bifurcation.

Transcendental Empiricism

Kant is said to be a transcendental idealist because he searched for the conditions of possible experience and placed them squarely upon the subject. For Kant, experience is possible because of the subjective conditions that subtend it, namely the forms of sensibility (space and time) and the categories (quantity, quality, relations such as causation, and modalities such as possibility and necessity). The categories determine objects of possible experience *a priori*, which allows us to distinguish empirical concepts (which are hypothetical and testable) from transcendent concepts (which are beyond experience and thus untestable). As is well known, Kant thus critically delimited reason from interrogating transcendent concepts like God, the self and its possible freedom, and the world as a whole because they are beyond possible experience. Kant's critical project was therefore *transcendental*, in the sense that he sought the conditions of possible experience to distinguish the transcendent from the empirical. But his critical project was also *idealist*, since those conditions rest squarely upon the subject (space and time are subjective forms of intuition, just as the categories form a transcendental logic of subjective experience, and we experience appearances and not things in themselves). While admiring Kant's transcendentalism, Deleuze searches for the conditions 'not of possible experience, but of real experience', because it is there 'we find the lived reality of a sub-representative domain'.[1] Transcendental empiricism is *transcendental* because it searches for conditions, but it is also *empirical* because the conditions it seeks are those for the production of new and real experience rather than for possible experience represented in advanced. The element of transcendental empiricism is thus live and eventual, in short, real experience. Deleuze's search for the conditions of real experience reveals his admiration of Bergson, whose method of intuition afforded a division of nature into natural articulations liberated from the representational structure of thought characteristic of realism and idealism. Deleuze writes:

> Intuition leads us to go beyond the state of experience toward the conditions of experience. But these conditions are neither general nor abstract. They are no broader than the conditioned: they are the conditions of real experience. Bergson speaks of going 'to seek experience at its source, or rather above that decisive turn, where, taking a bias in the direction of our utility, it becomes properly human experience.' Above the turn is precisely the point at which we finally discover differences in kind.[2]

Bergson's method of intuition established differences in kind between the pure past and the present, memory and perception. These differences in kind were viewed as the source of real experience, which takes place, as it were, within the durational warp in the continuum of the past and the present. This durational vantage enabled Bergson to critique the representational model and its method of dividing nature spatially, according to differences of degree, whereas an adequate concept of nature must account for differences in kind that subtend those differences of degree. The representational model accounts for memory in the same way that it accounts for perception – memory is just an accumulation of perceptions and so the difference between the two is merely a difference of quantity or degree. This representational method of distinguishing memory from perception thus produced a figure of memory as a storehouse of perceptual images. Bergson, on the other hand, uncovered differences in kind between memory and perception. This qualitative distinction between memory and perception constitutes a critique of representation: memory cannot be considered in terms of perception because the past cannot be considered in terms of the present.

In a similar fashion, Bergson encourages us to reevaluate the notion of possibility. Possible experience, as it is normally understood, is real experience that simply has not been actualised. When we consider possible experience, we merely conceive of present experience that has not yet taken place. But just as the past is not understandable in terms of the present, so the possible is not conceivable in terms of the actual. We habitually trace the possible on the outlines of the present, as if it were a 'present' time that is merely waiting to occur, as if the possible future were a crowd of 'presents' stuffed into some kind of ontological reservoir. But Bergson constantly reminds us that there is 'more', not less, in the negative thought of the possible (it is negative because it negates the actuality of the real in order to arrive at the possible): 'the possible is only the real with the addition of an act of mind which throws its image back into the past, once it has been enacted'.[3] When we think of the possible, we simply take the real and negate its actuality – and modelling the possible on the counters of the real is what constitutes so many muddles of the representational model because it conflates differences in kind with differences of degree. The figure of the possible is thus kindred to the spatial picture of memory as a storehouse of images – in both cases we abstractly represent the past in terms of the present. This figure of the possible cannot account for real experience anymore than the present can account for the past. In order to account for real experience, a new concept of time and becoming is required, a concept that moves beyond the determinations of the present. It is in this regard that Deleuze's transcendental empiricism takes the mantle from Bergson's intuitive analyses of nature.

The lynchpin of Deleuze's transcendental empiricism is the concept of difference in-itself, apart from identity. Basically, Deleuze finds that historically, difference has always been understood in terms of identity, as the difference between things (like the difference between two words, books, a table and a chair, etc.). By excavating a concept of difference apart from identity, Deleuze formulates a concept that harbours an escape route from the bifurcation of nature, producing conceptual resources for a new philosophy of nature. However, Deleuze develops his concept of difference through a reversal of

Platonism, so our examination of the concept will require something of a detour from our previous focus on nature.

Deleuze's reversal of Platonism is itself a provocative line of inquiry that has drawn diverse evaluations. Daniel Smith, for example, discovers a 'rejuvenated ... and even a completed Platonism'[4] in Deleuze's appropriation of the simulacrum as a concept of difference, whereas Miguel de Beistegui finds chiefly an anti-Platonism in the concept of the simulacrum.[5] Far from being incompatible evaluations, however, these divergences instead reveal the simulacrum as the hinge between the dual accomplishment of Deleuze's reversal of Platonism, because while Deleuze disparages the legacy of transcendence that followed Platonism, he also preserves essential dimensions of Platonism, rehabilitating them within a differential register. On our way to Deleuze's concept of difference in-itself we will examine how Aristotle and Plato manage difference and individuation, Nietzsche's overturning of Platonism, and the problem of rivalry in ancient Athens.

The Image of Thought

Deleuze characterises the image of thought as an intellectual architecture engineered to subordinate difference to identity. That is to say, the image of thought does not account for difference as it is *in-itself*, but rather through identity. Deleuze writes:

> There are four principal aspects to 'reason' in so far as it is the medium of representation: identity, in the form of the *undetermined* concept; analogy, in the relation between ultimate *determinable* concepts; opposition, in the relation between *determinations* within concepts; resemblance, in the *determined* object of the concept itself. These forms are like the four heads or the four shackles of mediation. Difference is 'mediated' to the extent that it is subjected to the fourfold root of identity, opposition, analogy and resemblance.[6]

Deleuze frequently considers Aristotle as a figure who thoroughly embodies this fourfold seizure of difference through his categories. The highest genera in Aristotle form the categories (substance, quantity, relatives, quality, etc.) under which specific differences can be classified. In this fashion, Aristotle divides the world in terms of *apparent joints*: the two-footed creature is separated from the four-footed, but the difference here is mediated by the common *resemblance* bestowed upon both creatures by their 'footedness'.

Difference and Representation

Furthermore, the animal species itself differs in terms of *opposition*, such as the opposition between the rational animal and the irrational animal. But difference here is held within the identity of the concept of animal. Moreover, although this generic identity in itself is undetermined, it acquires determination by specific differences that must be imported from outside of the genus. That is to say, specific differences such as 'rational' determine the undetermined concept 'animal' *externally*, in the sense that the concept 'animal' does not entail or determine 'rational' within its own concept. 'Rational', then, as a specific difference, is a determination arriving from *outside* of the genus 'animal'. Now, when we reach the highest genera the situation changes significantly, because the categories are too different to be said to differ in virtue of a commonality, and so the difference between the highest genera is said to be *analogous* to those specific differences. Aristotle famously wrote that 'there are many senses in which a thing may be said to "be", but they are related to one central point, one definite kind of thing, and are not homonymous'.[7] This feature of the *pros hen* equivocity of Being is commonly understood in terms of the argument Aristotle presents in order to support his claim that Being is not itself a genus: the highest genera cannot be accounted for by understanding Being as the highest genus, because that would render the highest genera as specific differences. The problem with understanding the categories as specific differences under the genus of Being is that those specific differences would cease to be enclosed within Being. This follows from the fact that genus is determined externally by specific differences, whereas being cannot be said to be determined by generic differences:

> genus is determinable only by specific difference from without; and the identity of the genus in relation to the species contrasts with the impossibility for Being of forming a similar identity in relation to the genera themselves. However, it is precisely the nature of the specific differences (the fact that they *are*) which grounds that impossibility, preventing generic differences from being related to being as if to a common genus (if being were a genus, its differences would be assimilable to specific differences, but then one could no longer say that they 'are', since a genus is not in itself attributed to its differences).[8]

In other words, the reason that one would attempt to understand being as a genus would be to gather the genera together under a commonality within which they could be said to differ from each other by reference to that commonality, but in doing so one would dismantle the possibility of saying that the genera *are* (different) because a genus is not attributable to its own differences. Again, the genus 'animal' is not comprehended in the specific difference of 'rational' or 'man', as the specific differentiae of 'animal' fall outside of the concept. If Being were a genus, then specific differences would fall outside of it, thus rendering those differences non-being. This, of course, seems absurd to Aristotle.

Now, this method of handling difference does not account for difference in-itself because difference is always accounted for in relation to identity in the form of (1) the concept (the genus), (2) analogy (on the one hand, in the relation between the highest genera and being, and on the other, in the relation between specific differences and genera), (3) opposition (that determines the species within a genus), and (4) resemblance (in the compared objects). Difference is thus held in the gridlock of the categories, the conceptual apparatus of representation. But difference was not always represented in this fashion. As we shall see, Plato managed difference in a remarkably different way from Aristotle.

Individuation in Plato and Aristotle: Selection vs. Mediation

MEDIATED INDIVIDUATION IN ARISTOTLE'S METAPHYSICS

A chief problem of Aristotle's *Metaphysics* is how to account for individuation. The problem revolves around identity and difference: what makes an individual the individual that it is (identity) and what makes it distinct from other individuals (difference). Aristotle's solution to this problem was to provide a method of classification, which accounted for the individual largely in terms of 'whatness' (or *quiddity* for the scholastics). What an individual is can be determined through the categories, genus and species, so the human individual is determined through the specific difference of 'rational' underneath the genus of 'animal'. But this method of individuation accounts for difference in terms of identity (again, the difference between a 'two-footed' creature and a 'four-footed' creature is managed through the identity in the concept of 'footedness').[9] Smith notes that Aristotle must 'relate Being to particular beings' through the concept of individuation, but that the concept '*cannot say what constitutes their individuality*: it retains in the particular (the individual) only what conforms to the general (the concept)'.[10] The critical argument here is that Aristotle inadequately accounts for differences by *representing*

Difference and Representation

given individuals through generic and specific classifications. Aristotle is very clear about his procedure of beginning with the given in *Metaphysics Z.17*:

> The object of the inquiry is most overlooked where one term is not expressly predicated of another (e.g. when we inquire why man is), because we do not distinguish and do not say definitely 'why do these parts form this whole'? ... *Since we must know the existence of the thing and it must be given*, clearly the question is *why* the matter is some individual thing, e.g. why are these materials a house? (1041a33)[11]

Thus he takes the given, the *haecceity* or 'thisness' of 'this man' or 'this house', and attempts to explain its individuation in terms of *quiddity*, or 'whatness', thereby basing the principle of individuation on this or that element of an already given individual. Looking at the given individual in terms of 'whatness', however, puts the cart before the horse since the only differences that can be accounted for are those that can be represented, i.e. those differences that can be mediated through generic and specific differentia. The sheer existence of the platypus, a duck-billed and beaver-tailed egg-laying mammal, presents a serious problem to the Aristotelian naturalist (in fact, British scientists in the late eighteenth century, who also wanted to preserve the integrity of genera/species classifications, believed the fantastic specimens to be the creations of talented Chinese taxidermists).[12] In either case, determining the platypus in terms of the 'whatness' provided by the categories is a frustrating exercise. And for Deleuze, this marks the chief problem with Aristotelian individuation: it restrains difference through the gridlock of the categories, which distribute differences through conceptual identities in the service of determining 'whatness'.

IMMEDIATE SELECTION IN PLATO

Plato, on the other hand, did not mediate difference through categorical concepts, through middle terms, but rather held that things *immediately* participated in their form or Idea: the 'just' man was 'just' by virtue of his participation in the form of justice, and the 'house' was a 'house' by virtue of its participation in the form of 'house'. There was no necessity in Plato to mediate differences between beings through the categories, genus and specific differentiae. As Deleuze notes, this treatment of difference was the subject of Aristotle's famous criticism of Plato, namely, the fact that difference 'operates without mediation, without middle term or reason; it acts in the immediate and is inspired by the Ideas rather than by the requirements of a concept in general'.[13] This criticism is levelled against Plato in *Metaphysics* Z.6, where Aristotle considered Platonic forms as the essences of things severed from the actual things, facilitating a relation between thing and form that could be characterised as 'participation'. In Z.6 Aristotle disputes this by providing an argument along the lines of the so-called 'third-man argument': if the (formal) essence of good is not the good thing, then one needs a third thing by which both are said to be good, and so on. Aristotle writes: 'The absurdity of the separation [between essence and thing] would appear also if one were to assign a name to each of the essences; for there would be another essence besides the original one, e.g. to the essence of horse there will belong a second essence' (1031b28). If Equus, Secretariat and Mister Ed are all horses, it is because they participate in the formal essence of horseness. But if horseness is not itself Equus, Secretariat or Mister Ed, then one requires another term in order to explain how Equus, Secretariat, Mister Ed and horseness are all 'horsenesses'. The search for an ultimate equestrian category thus gallops into an infinite regress. Notwithstanding, Deleuze appreciates the fact that in Plato the dialectic of difference operates without mediation between the form or the Idea and the thing. Taking Aristotle's criticism as grist for Plato's mill, Deleuze asks:

> Is this not its strength from the point of view of the Idea? ... Division is not the inverse of a 'generalisation'; it is not a determination of species. It is in no way a method of determining species, but one of *selection*. It is not a question of dividing a determinate genus into definite species, but of dividing a confused species into pure lines of descent, or of selecting a pure line from material which is not.[14]

At once we note that while Aristotle focuses on how to mediate the *separation* of the Idea from the being, Deleuze focuses on the *function* of division in order to uncover the Idea as a principle of filiation. While mediation determines individuals specifically and generically, the Idea selects individual claimants according to their lines of participatory descent in a kind of election by formal participation.[15] The problem for Plato and mimetic division was to *authenticate* and select the pure from the mixture, the philosopher from among the competing sophists, whereas in Aristotle the problem changes into a question of *knowledge*, a question of determining individuals through a classification into natural kinds, genera and specific differences. Thus Deleuze finds in Plato 'nothing in common with the concerns of Aristotle: it is a question not of identifying but of authenticating. The one problem which recurs throughout Plato's philosophy is the problem of measuring rivals and selecting claimants', and Plato selects the philosopher by 'making the difference' (*faire la différence*) between the philosopher and the sophist.[16] The significance of Deleuze's endorsement of the Platonic Idea over and against Aristotelian categories lies in the fact that Deleuze senses, within Platonism itself, the problem of having to make the difference in order to measure rivals, a problem which can no longer be detected in Aristotle. Where Aristotle mediates, Plato selects: gold in Aristotle is something to be determined in a classification of metals that includes silver and bronze, whereas for Plato gold is something that must be selected as authentic or inauthentic, true or false gold. In Plato one senses the contest that *precedes* the measure of rivals, thus the dialectic deployed to discern the true from the counterfeit operates in the 'depths of the immediate', so it is 'a dialectic of the immediate' as opposed to a categorical mediation.[17]

THE ROLE OF MYTH IN PLATONIC SELECTION

A crucial element of this dialectic of the immediate is myth. Deleuze reminds us that in Plato the *logos* unfolds alongside *mythos*, because whenever there is a question of selecting claimants, a question of dividing and classifying rivals, a myth is invoked. In the *Statesman*, for example, we find the mythological image of an ancient God who ruled man and world, and only he, strictly speaking, deserves the name of shepherd-king. However, after invoking the mythic shepherd-king as a standard, claimants can be measured according to an order of participation in that ancient model; subsequently it becomes possible to distinguish between the most proximate or 'truest' statesman and parents, servants, auxiliaries and mere counterfeits in terms of their likeness to the ancient model.[18] Similarly, whenever the question of distinguishing between different 'madnesses' is broached in the *Phaedrus*, Plato invokes the myth of metempsychosis, the circulation of souls before reincarnation – only by reference to what the lover-soul has contemplated in its mythological prehistory can we discriminate between true and false lovers.[19] Finally, regarding the third important text concerning division, the *Sophist*, Deleuze finds that the lack of myth in the dialogue serves as a means to discover the sophist: 'the point is that in this text, by a paradoxical utilization of the method, a counter-utilization, Plato proposes to isolate the false claimant *par excellence*, the one who lays claim to everything without any right: the "sophist"'.[20] The sophist, then, neglects the essential mythological dimension of dialectic and thereby reveals his illegitimacy. In fact, it is possible to argue that this dialectical procedure was performatively schematised in the *Phaedo*, because whenever things are not 'sure' or 'clear' (*saphes*) to the interlocutors, Socrates insists that they must 'look through the stories they tell' (*diaskopein te kai mythologein*) in order to arrive at the account (*logos*) hidden within them, as the soul is hidden within the body.[21] In the *Phaedo*, it is only through the provisional depths invoked by myths that we are able to measure, grade and select the true and best account of the fate of the soul after death. The point throughout is that the *logos* is inadequate by itself – a different kind of speaking, a *mythologein*, is necessary for inquiries to exhibit any truth. But myth here does not erect a *transcendent* Idea over the apparent claimants; rather, 'myth', Deleuze tells us, 'constructs the *immanent* model or the foundation-test according to which the pretenders should be judged'.[22] The depths of the Platonic dialectic cannot be explored in terms of a duality between the immanent and the transcendent or the sensible and the intelligible. Mythological histories function as depths from within which the *logos* is deployed in order to make a selection among rivals; although, crucially, these depths remain beyond the register of the *logos* and representation.

In this fashion, Deleuze reveals the theory of forms as a solution to the problem of selection between rival claimants. But as we will see in the following section, Nietzsche's studies of Plato had already pointed toward an agonistic motivation behind the Ideas, i.e. the fact that the Ideas are deployed in the service of establishing of a criterion by which one can *decide* between rivals as if it were a question of *deciphering* the true from the false, the philosopher from the sophist. Advancing Nietzsche's work on Plato, Deleuze's most original contribution to the history of Platonism lies elsewhere; namely, in a reevaluation of the simulacra, which allows him to discover that the problem of difference had not yet been completely annulled in Plato, which in turn allows him to reevaluate the Idea in a different manner from Nietzsche. Ultimately, the problem of difference and the concept of the simulacrum will allow us to reconceive the problem of nature disabused by the traditional categories of bifurcation – the subject, the object, and the supposed correspondence between them

Nietzsche and the Overturning of Platonism

In a celebrated section of *Twilight of the Idols* subtitled 'How the "Real World" at last Became a Fable', Nietzsche traces the history of the hierarchical division of being into true and merely apparent realms, a bifurcation of the world that had its origin in Platonism. If the real world no longer holds its persuasion, then 'what world is left?' Nietzsche asks, 'the apparent world perhaps? ... But no! [W]ith the real world we have also abolished the apparent world!'[23] This is commonly understood to be a wholesale condemnation of Platonism as a 'two-world' theory of being – a world of eternal essences set over and above a world of changing appearances. Eager to correct this common over-simplification, Deleuze reminds us that 'the dual denunciation of essences and appearances dates back to Hegel or, better yet, to Kant. It is doubtful that Nietzsche meant the same thing.'[24] As we shall see, a simple dichotomy between essences and appearances or between transcendence and immanence cannot adequately frame Nietzsche's or Deleuze's relation to Plato. Deleuze will obliquely advance Nietzsche's provocative thesis: 'The task of modern philosophy has been defined: to reverse Platonism.'[25] But only obliquely, as he maintains the caveat: 'That this reversal [*renversement*] should conserve many Platonic characteristics is not only inevitable but desirable.'[26] Since Deleuze frames his own 'reversal' of Platonism in the context of Nietzsche's 'overturning', it behooves us to examine the latter before approaching the former.

Nietzsche on Language

In order to examine Nietzsche's overturning of Platonism in *Twilight of the Idols*, a brief analysis of another of Nietzsche's texts, 'On Truth and Lying in an Extra-Moral Sense', proves to be of service. This latter work can be characterised as an intervention between words and truth, or in the Platonic formulation, between *logos* and origin.[27] Nietzsche writes: 'The various languages placed side by side show that with words it is never a question of truth, never a question of adequate expression.'[28] His argument outlines a genetic account of language in terms of the creation of metaphors that aims to overturn the imitation of Platonism. For Nietzsche, words are metaphors (literally 'after-bearers' – *meta* 'after' + *phero* 'to bear' or 'to carry') that are merely imputed to correspond to an original world, but the original world to which they are said to correspond is itself a specious addition to words, an after-effect. Nietzsche's argument against the correspondence theory of truth lies in the surreptitious production of the world that words are said to 'correspond' to – such a world is merely a function of putting what 'comes at the end', namely, the idea of an original world, 'at the beginning as the beginning', as if the product was a cause.[29] 'What is a word? It is the copy in sound of a nerve stimulus. But the further inference from the nerve stimulus to a cause outside of us is already the result of a false and unjustifiable application of the principle of sufficient reason.'[30] In this fashion, Nietzsche argues that we cannot simply postulate a world of causes behind a world of words and appearances. And this constitutes Nietzsche's chief complaint against Platonism, because while Plato did not postulate a 'thing' outside the sensory world that words might express; he nonetheless set the history of philosophy in that direction by positing an entire community of original forms through which subsequent 'copies' could be gathered and unified. And for Nietzsche, it is this postulate that is the most deceptive:

> Every word instantly becomes a concept precisely insofar as it is not supposed to serve as a reminder of the unique and entirely individual original experience to which it owes its origin; but rather, a word becomes a concept insofar as it simultaneously has to fit countless more or less similar cases – which means, purely and simply, cases which are *never equal* and thus *altogether unequal*. Every concept arises from the *equation of unequal things*.[31]

Moreover, it is through this equating procedure, a procedure of evaluating differences through the identity of the concept, that we are able to postulate a 'real' world populated with such original identities:

> This awakens the idea that, in addition to the leaves, there exists in nature the 'leaf': the original model according to which all the leaves were perhaps woven, sketched, measured . . . – but by incompetent hands, so that no specimen has turned out to be a correct, trustworthy, and faithful likeness of the original model.[32]

It is not merely the possibility for adequation or fitness of words and things that is relinquished here, but rather our entire correspondence to the origin is severed, and our metaphors are left to have meaning only in a situation involving a complex of other metaphors, which

Difference and Representation

Nietzsche refers to as a 'mobile army'. There is no primal leaf, no original leaf. Moreover, the significance of forgetfulness, as the condition for forming concepts, applies to Aristotle's categories as much as to the Ideas in Platonism: we forget the difference between things and so arrive at consolidated metaphors (concepts) posited but not real. In this sense, we abstract from differences in order to equate sensations into identities or neutralised objects, so that we eventually posit a unity to which all differences correspond, i.e. a primal leaf, the Idea, concept, or species of leaf. Regarding Platonism, Nietzsche's worry is that, like Aristotelian categories, the Idea as an origin functions as a homogenising mechanism that determines the phenomenal leaves *in advance* – the leaves are apprehended as leaves only *after* the primordial leaf (the word for Platonic participation is *metechein* – literally, 'to come after').[33] The Idea is thus understood as an abstraction that equalises and neutralises differences. This brings us to the most fascinating point here, because the intellect, which functions through the deception of forgetting, does not spare anything, even itself: the intellect uses its deceptive power against itself by forgetting this very procedure, forgetting its own process of forgetting. Forgetting is the ultimate neutralisation of the creative advance of difference.

Since forgetting is a process of abstracting from differences in order to arrive at consolidated identities, we may profitably compare Nietzschean forgetting to the procedure of abstraction operative in Whitehead's fallacy of misplaced concreteness. Whitehead argues that we commit ourselves to fallacy when we abstract or disconnect an object from its concrete involvement within process (or the passage of nature) and subsequently (and fallaciously) 'misplace' the abstraction underneath the concrete as its substratum. The fundamental ground of nature for Whitehead is passage or becoming, and not individual entities or beings. Thus, Whitehead's fallacy of misplaced concreteness, like Nietzsche's reactive forgetting, is a critique of reification. The fallacy is thereby aimed at a certain type of discursive knowledge that surreptitiously recasts reified entities as the ground of natural process, and so 'what is a mere procedure of mind in the translation of sense-awareness into discursive knowledge' is subsequently 'transmuted into a fundamental character of nature'.[34] However, while Whitehead did not consider the fallacy as 'a vice necessary to the intellectual apprehension of nature', but rather as 'merely the accidental error of mistaking the abstract for the concrete', Nietzsche saw forgetfulness as constitutive of discursive knowledge itself.³⁵ In this fashion, Nietzsche's philosophical intervention is aimed directly at discursive knowledge *tout court*, and his agenda is to overturn Platonism by overturning the image of intelligibility itself. If concepts are formed through a reifying process that forgets differences, that is to say, through a process of unconscious deception, then the 'real world' of intelligibility is founded upon such an unconscious deception, and thus becomes a fable.

THE SOCIAL AND POLITICAL CONDITIONS OF TRUTH AND FALSITY

Truth, then, becomes merely a product of linguistic legislation, in the sense that proper or 'truthful' ways of using words are simply *conventional designations*.³⁶ But that means that truth and falsity as determinations are grounded upon socio-political conditions. In war, for example, it is best to deceive the opponent, to flout conventions and expectations, to play upon what seems, but is not. But in peace, the liar (who uses words in ways contrary to their conventional designations) is exiled from the city, because, to use Nietzsche's example, he says 'I am rich' when the conventional designation would be 'poor'.³⁷ The true here is not defined by some perfect adequation with, or correspondence to, reality, but rather through its *function* in a socio-political context. Notwithstanding, this socio-political condition of discursive knowledge or truth is forgotten, and so metaphors are taken to designate a reality of things-in-themselves or essences, and no one recalls that the concept is 'merely the residue of a metaphor'.³⁸ But if the socio-political artificiality of concepts is forgotten, the creative personality that employs language in its deceptive capacity (albeit in all its creative innocence) nonetheless remains to be judged according to the linguistic legislation that henceforth has become an imperative. In this fashion, the play of dissimulation that reigns in agonistic encounters is invalidated through the establishment of the concept, and man forgets that he is an essentially creative subject (a creator of metaphors) in order to live in the 'repose, security, and consistency' of the state.³⁹ Ultimately, in so far as the socio-political conditions of truth that determined the validity of designations is forgotten, a hierarchical state emerges, inscribed within words themselves, that subsequently establishes a criterion by which to evaluate claims, thereby transforming the playing field between rivals, eliminating the creative form of the contest itself.

Nietzsche uncovers, then, a *reactive politics of the Idea* in Platonism, a politics that emerges as the greatest accomplishment of reactive forgetting: an intelligible original according to which sensible manifestations are to be measured. The establishment of an intelligible original form of something, however, demands that sensible phenomena take on the status of 'copies'. Henceforth, phenomena can be more or less faithful copies of the original, they could simply 'seem' to be something while truly being something else; and the original form can be invoked in order to judge between more or less faithful copies, to judge between true and false copies, to judge, in the final analysis, between Socrates and the sophists.

The Problem of Rivalry

In 'Homer's Contest', Nietzsche reminds us that Hesiod's *Works and Days* presents two Eris-goddesses (goddesses of envy or strife) – a first, older Eris goddess promotes war and cruel feuding, and a second, younger one promotes vigorous competition. Hesiod notes that this latter Eris is good for men because she 'drives even the unskilled man to work; and if someone who lacks property sees someone else who is rich, he likewise hurries off to sow and plant and set his house in order . . . even the potters harbor grudges against potters, carpenters against carpenters'.[40] While the first and older Eris brings men to life or death struggle in war, the second younger one brings men to healthy envy that goads them to compete, that vitalises agonistic encounters. Holding grudges, even excessive envy, seems to be not only a defining characteristic of the Hellenic ethic but a praiseworthy one, in the pious sense of praise for the younger Eris goddess.

THE CRISIS OF SOVEREIGNTY: DEMOCRACY AND SPEECH IN THE GREEK *POLIS*

Jean-Pierre Vernant, in his work on the Greek *polis*, argues that rivalries such as those between Socrates and the sophists were an essential dimension of the Athenian city.[41] The tool for such rivalries was discourse (*logos*): speech was both the instrument of combat for the engaged rivals, but also the medium of judgement for the competition itself. Vernant writes: 'speech was no longer the ritual word, the precise formula, but open debate, discussion, argument. It presupposed a public to which it was addressed, as to a judge whose ruling could not be appealed, who decided with hands upraised between the two parties who came before him.'[42] The publicising of the authority of language brought the divinely invested authority of the priest down to the level of the popular measurement of persuasive speech, down to the horizontal field of the public and agonistic *agora*:

> The *polis* was seen as a homogeneous whole, without hierarchy, without rank, without differentiation. *Arche* was no longer concentrated in a single figure at the apex of the social structure, but was distributed equally throughout the entire realm of public life, in that common space where the city had its center, its *meson*.[43]

Vernant is not merely relating the architectural layout of ancient Athens to the social phenomena of a democratic public debate. Rather, here we have an instance of how a social space is technically arranged so as to conduct a series of democratic events.[44] One may even say that the popular use and conception of speech is inscribed within the very architecture of the social space and time of the *polis*. Many commentators have noted that this architectural distribution of the city, which was indebted to Cleisthenes' democratic reforms in 507–506 bce, was governed by a principle of *isonomia* (literally, 'equality of law'). And in addition to the secular reorganisation of space that placed the agora at the centre of the *polis*, Cleisthenes' reforms inaugurated a political calendar (in terms of periodically assembled councils of citizens called prytaneis) that arranged time differently from the religious calendar (in terms of lunar phases).[45] Here we have an arrangement of civic space and events that facilitates democratic rivalries. A century before Plato, Hippodamus had radically equated the urban order with the isonomic political order, installing regular gridirons of streets intersecting at right angles until the town became 'indefinitely extensible' and 'risked dissolving into infinite space', to the degree that Hippodamus felt the need to set 'a limit to the number of citizens' and 'impose a finite terminus to the urban space'.[46] Thus, the horizontal redistribution of space (as opposed to the vertical distribution of a temple or palace) threatened an infinite democratic mob. In any case, by the time of Plato, the principle of *isonomia* was firmly installed within the spatial and temporal architecture of the *polis*. So Vidal-Naquet tells us that the agora was a 'place where problems of general interest were argued and where power was no longer located in the palace but in the center, *es meson*. It is "in the center" that the orator stands, the one who is supposed to speak in the interest of all.'[47] Clearly, many of Plato's dialogues stage this contest between Socrates and his rivals, and from within finds an agonistic motive driving the theory of forms that separates Socrates from his rivals. But this dynamic of rivalry can only take place within a certain form of society; that is to say, the element of rivalry implies a certain social order. In order to have genuine rivalries, one needs a democracy of free men, a social arrangement where citizens are encouraged to challenge each other. In other words, there can only be a genuine contest among citizens if the citizens inhabit a social order that facilitates such encounters – an isonomic social order, for example. Rivalry, after all, implies a right to enter into contest or a right to lay claim to something (and so a social order where many may rightly claim, 'I am the shepherd of men', 'I am the true lover'). This 'crisis of sovereignty', Vernant observes, implies an equality among men: 'all rivalry, all *eris* presupposes a relationship of equality: competition can take place only among peers'.[48] But if speech is the medium of competition, then changes in the competition will be accompanied by changes in the way the claimants speak.

Precisely such a change in the mode of speaking arrives with the Platonic Ideas. However, while the Idea begins in the agonistic milieu of the agora, Nietzsche chronicles how it gradually grows more and more distant, eventually facilitating a mode of speaking of a world 'unattainable for the moment, but promised to the wise, the pious, the virtuous man ... it becomes Christian', and continues in this direction until it grows 'sublime, pale, northerly, Königsbergian'.[49] The important point to note here is that Nietzsche's overturning of Platonism has less to do with Plato than with the historical unfolding of the Idea – the Idea was only *too* victorious, and the contest has all but disappeared. It is precisely in this regard that Deleuze takes up the mantle from Nietzsche – to find the problematic conditions from whence the Idea arose, and to recover the Idea as it operated within its natural habitat, as it were. The Idea may conclude in some elite transcendent realm, but it was born in the deafening roar of the agora.

Deleuze and the 'Reversal' of Platonism

PLATO'S SOLUTION TO THE PROBLEM OF RIVALRY: MODEL, COPY, SIMULACRUM

In *Difference and Repetition*, Deleuze articulates how the Platonic division of being that separates intelligible Ideas and sensible images has a primarily *elective*, and not ontological, motivation:

> The *true* Platonic distinction lies elsewhere: it is of another nature, *not between the original and the image but between two kinds of images* [*idoles*], of which copies [*eikones*] are only the first kind, the other being simulacra [*phantasmata*] ... The function of the notion of the model is not to oppose the world of images in its entirety but to select the good images, the icons which resemble from within, and eliminate the bad images or simulacra.[50]

Similarly, in *The Logic of Sense*, Deleuze concludes that the 'great manifest duality of Idea and image is present only in this goal: to assure the latent distinction between the two sorts of images and to give a concrete criterion'.[51] The point throughout is the fact that Plato was presented with a problem of evaluating the sophists, with whom Socrates wrestled for distinction. His resolution to that problem came in the form of a conceptual framework that distinguished copies from simulacra in relation to the Idea. It is as if everything in Plato were split in two: in the *Gorgias*, rhetoric is disparaged as empty flattery and persuasion, while in the *Phaedrus* it holds an affinity with truth; the *Statesman* separates the true shepherd of men from the false; the *Phaedrus* attempts to distinguish between true and false lovers; and, of course, the sophist must always be distinguished from the philosopher. But wherever these two powers of images appear, the establishment of Ideas or forms provides a way to evaluate between true copies (*eikones*) and false copies (*phantasmata*) in terms of their likeness to the form, and so the Idea thereby serves as an arbiter between true and false claimants to the truth, that is to say, between Socrates and his rivals. Now, one might be tempted to understand the Platonic manoeuvre here in terms of immanence and transcendence. On this reading, Platonism's erection of transcendent Ideas pulls the contest, as it were, from underneath Socrates' rivals: the agonistic play is no longer staged on the horizontal agora, but the contestants are rather vertically rearranged on a scale of transcendence. But as Nietzsche shows, the transcendent hierarchy of an intelligible world over a sensible world belongs to a later history of Platonism; the dialogues themselves present a different configuration of the Ideas. Of course, this is no place to present an argument regarding forms and participation in Plato, but Deleuze's reversal of Platonism is oversimplified by the claim that the Idea in Plato is simply transcendent whereas in Deleuze the Idea is purely immanent. Deleuze finds the Platonic distinction to function within an immanent field – not between the sensible and the intelligible, but between two kinds of images, regarding which

Difference and Representation

the Idea serves as an 'immanent model' emerging from myth.⁵² Somewhat tangentially, but importantly: Nietzsche and Deleuze do not subscribe to a two-world view of Plato, and neither does the Idea require a two-world interpretation in order to disqualify Socrates' rivals. In fact, the mythological ground of the deployment of the Ideas is persuasive precisely because it is capable of drawing a line of descent that can manifestly bear upon the rival claimants. It is only because the Ideas are immanent that they can render Socrates' rivals false copies, far removed from the Idea, and so the contest may have a victor – Socrates.⁵³

But the primary reason that Deleuze disapproves of the standard appraisal of Nietzsche's overturning of Platonism, as the simple abolition of the intelligible as well as the sensible, is because such a caricature 'has the disadvantage of being abstract; it leaves the motivation of Platonism in the shadows'.⁵⁴ Without locating the problems that condition philosophy, philosophy remains abstract – and there is nothing abstract in Plato, or Deleuze for that matter. To reverse Platonism, for both Nietzsche and Deleuze, means to retrace it back to its concrete and problematic ground; to reverse Platonism is to engage a genealogy of the Idea. If Plato introduces the Idea into philosophy, the question becomes: why Plato in Athens? What problem was Plato responding to by creating the concept of the Idea? Far from an airy realm of intelligible Ideas, Deleuze senses a Heraclitan world still growling within the dialogues, so from Nietzsche's exposure of the deceptive character of the Idea, he embarks upon his own positive reversal of Platonism that preserves and rehabilitates the Idea in a differential register. Deleuze will thereby recover the Idea from the facile two-world view of Platonism while illuminating and redirecting Nietzsche's critical engagement with Platonism.

As previously mentioned, in Aristotle, a concept of difference in-itself was irrevocably annulled by identity through the mediation of the categories – there was no way of conceiving of difference apart from identity. In Plato, however, the problem of difference is still discernible, one still feels the contest in the dialogues. Thus, Deleuze finds that 'the issue is still in doubt':

> mediation has not yet found its ready-made movement. *The Idea is not yet the concept of an object which submits the world to the requirements of representation, but rather a brute presence which can be invoked in the world only in function of that which is not 'representable' in things.* The Idea has therefore not yet chosen to relate difference to the identity of a

concept in general: it has not given up hope of finding a pure concept of difference in itself.[55]

Among other things, this means that the Ideas in Plato have not yet bifurcated the world representationally into a realm of eternal Forms by which the changing manifestations of the sensible realm are to be judged and evaluated. The mythological element by which the dialectic proceeds confirms that Platonism has not yet submitted the world to the requirements of representation: one would have to extricate the Ideas from their myth-telling (*mythologein*) ground in order to reconstitute them as a set of identities by which differences are to be evaluated. Since Plato had not subjugated difference to mediation (the linchpin of representation), difference in-itself remains discernible in the dialogues, but as a threat that requires order and control. Thus, by discovering the agonistic motivation behind the deployment of the Ideas, Deleuze reveals that Ideas, far from being elements of representation, are a function of what is *not* representable in things: the presence of the Ideas testifies to the fact that one *could not* discriminate between rivals, that there was no standard measure by which to grade claimants, in short, that rivalry presented a genuine problem, that is, a problem that did not have a clear solution. Thus, far from being a concept that bifurcated the world into appearances and essences, Deleuze's genealogy of Platonism reveals the Idea as a function of the problem of selecting rivals.

The Ideas were deployed, then, in order to *make a difference* between true copies (*eikones*) and false copies or simulacra (*phantasmata*). But this entails that the simulacrum is a necessary term of contrast for the process of division, as one does not significantly distinguish a true copy from a true copy. Now, the copy can be said to imitate in so far as it stands against a model. But what if we examine the simulacrum by itself, apart from both model and copy? Such an examination would yield only pure difference: not being copies or models themselves, and without reference to something which they would imitate, we cannot even say that simulacra are different 'from' anything, which is to say they are disparate in themselves – they are difference and only difference, an *internalised difference*. Within the simulacrum, therefore, Deleuze finds 'a becoming-mad, or a becoming unlimited', a figure of *apeiron* (the boundless or the 'unlimited') that cannot be located or determined apart from the imposition of an external limit, 'a becoming always other, a becoming subversive of the depths, able to evade the equal, the limit,

Difference and Representation

the Same, or the Similar: always more and less at once, but never equal'.[56] How does one identify the unlimited by itself, the different by itself? Apart from identity, the simulacra can only show up as so many masks, and masks behind masks.[57]

We are now in a position to briefly examine how the concept of difference is so well suited to a philosophy of nature based on time (instead of space): difference understood as becoming renders a picture of time as intensive, rather than a line upon which space passes in successive and homogeneous instants. The simulacrum, as a figure of pure difference, allows us to conceive of a concept of becoming apart from identity, that is, a concept of pure becoming or pure change. This would be a concept of becoming where there is no 'thing' that becomes something else, but only a becoming without any identifiable beginning or end. In *Nietzsche and Philosophy*, Deleuze treats pure becoming through Plato's second hypothesis in the *Parmenides*: 'Plato said that if everything that becomes can never avoid the present then, as soon as it is there, it ceases to become and is then what it was in the process of becoming.'[58] In other words, if the present (*parousia*) is taken as being or substance (*ousia*), then it is opposed to becoming (coming-to-be), because it *is* (present). However, if the present is taken as being, then the present, stripped of becoming, would never pass because in order to pass it would need another present to push it, as it were, into the past. The problem with identifying the present as being lies in the fact that it requires a movement of being, a concatenation of presents that successively force each other in and out of presence. Thus, the present cannot be understood as being because being is a static figure, self-same and eternal, and if the present adopted such a figure it would never 'become'. Moreover, if becoming itself would come to *be*, then becoming would cease to be becoming, that is to say, it would cease 'to become' in simply being (present). In this fashion, becoming *as such* would never come to be, and neither would it end (in coming-to-be the present). Consequently, the only way to think of *pure becoming* is to think of a difference that repeats itself without end, that never concludes in an identity but rather is purely productive, a becoming with neither *arche* nor *telos*. It would be like a figure of a tunnel without entrance or exit, a figure of pure passage, transformation and creativity.

Similarly, the simulacrum operates in a field that repels and excludes any origin, that is to say, it operates in the register of difference in-itself, wherein difference escapes mediation and is thereby related immediately to difference. Thus, the simulacrum reveals difference beyond origin, not an originary difference. This is precisely the meaning of the *differential* for Deleuze: 'systems in which different relates to different through difference itself are systems of simulacra. Such systems are intensive.'[59] It is in this capacity that the simulacra reveal an intensive or *transcendental* field beyond representation. Nonetheless, the simulacra are necessary for representation, as the false 'copy' needs to be distinguished from the true. Thus, the simulacrum comes to form a *necessary yet subversive* concept vis-à-vis representation, the structure of the model and the copy. In *Difference and Repetition*, Deleuze finds that, 'simulacra provide the means of challenging both the notion of the copy and that of the model ... the model collapses into difference, while the copies disperse into the dissimilitude of the series which they interiorise, such that one can never say that the one is a copy and the other a model'.[60] And in *The Logic of Sense*, Deleuze muses about the moment in which Plato, in acknowledging the simulacrum, must have recognised this challenge:

> as a consequence of searching in the direction of the simulacrum and of leaning over its abyss, Plato discovers, in the flash of an instant, that the simulacrum is not simply a false copy, but that it places in question the very notations of copy and model ... Was it not Plato himself who pointed out the direction for the reversal of Platonism?[61]

We may imagine Deleuze to have experienced a similar flash of discovery when he saw that in Plato the issue is still in doubt – difference in-itself has not been categorically annulled by identity. Here we find the chief virtue of the simulacra for a philosophy of difference. The creation of a concept of difference in-itself requires the dismantling of representation, or the structure of original model and copy. And the simulacrum reveals itself precisely in this critical or subversive role: while the simulacrum is *necessary* for Plato to distinguish the true from the false claimant, it is nonetheless a *subversive* concept because it at once presents an internalised difference which dismantles the notion of original identities, repelling the capture of both model and copy, and, finally, bringing into question the legitimacy of both. 'For Socrates distinguishes himself from the Sophist', Deleuze writes, 'but the Sophist does not distinguish himself from Socrates, placing the legitimacy of such a distinction in question. Twilight of the *icônes*.'[62]

Nietzsche's evaluation of the Idea found chiefly a reactive politics of identity: the Idea served an equalising function to annul differences,

Difference and Representation

ultimately suggesting an origin to the world, qualifying appearances as so many imperfect copies. But Deleuze finds, within the problem of rivalry, an *active politics of the simulacra*. By presenting an internal division at the core of Platonic division, the simulacra reveal unmediated difference not simply as a remainder, but as an essential dimension of the identification of copy and model. Plato, however, never gave the simulacrum its own concept, and instead chose to manage the constitutive disparity of the simulacra through the framework of model and copy, thereby curbing the subversive potential of the simulacrum, that is, its potential to outstrip the rank from not only the copy but also the model. Deleuze, on the other hand, affirms this subversive potential, and gives the simulacrum its own concept, a concept of difference in-itself. In a mutiny of masks, as it were, the simulacra assert a problematic register wherein copies *can no longer be differentiated* from models, which is precisely why the concept of the simulacrum becomes the emblematic figure of difference in Deleuze's reversal of Platonism – it is the swansong of identity.

DELEUZE ON THE IDEA IN PLATO

The role of Plato in Deleuze's own theory of Ideas is often disregarded in favour of other thinkers, such as Albert Lautman and Kant.[63] Deleuze himself encourages this neglect with statements such as, 'the poisoned gift of Platonism is to have introduced transcendence into philosophy, to have given transcendence a plausible philosophical meaning (the triumph of the judgment of God)'.[64] This has tempted some commentators to dismiss Plato as merely an instance of transcendence, another representational theory. But what does transcendence mean here? Surely it does not designate a transcendent world. So we find that even in his most critical and summary comments about Platonism, Deleuze is careful to avoid such dismissive simplifications: in the short 1992 note that briefly treats the 'poisoned gift of Platonism', for example, Deleuze distinguishes between the peculiar type of transcendence of the Idea and other types of transcendence (imperial, mythical). Plato's transcendence is distinct because it is 'situated *within* the field of immanence itself. This is the meaning of the theory of Ideas.'[65] Mythological histories, of course, play a role here, but they are invoked in Plato in order to provide a ground for the Ideas, which are the principles of filiation that can decide which is the true lover, the true friend, the true shepherd of men. The function of the Idea in Plato was to install a ground where none was given: emerging from a mythological history which came 'first', the Idea was invoked so that claimants could participate second-hand, third-hand, etc. It is in this regard that the Idea is said to introduce transcendence into the immanent problem of rivalry. But this is a peculiar type of transcendence – one that remains within its immanent problematic field. In fact, it is difficult to see how the simulacra would even emerge in a social order characterised by a different kind of transcendence, such as, for example, that of the pre-Homeric Mycenaean world.

What little we know of Mycenaean Greece renders it unfathomably different from even Homeric worlds, outstripping any possibility of comparison with Plato's Athens.[66] A title frequently used in that period, *wa-na-ka* (*wanax*), refers to a figure that might be called a god, priest or king (Vidal-Naquet claims the term will always be mistranslated), who fused together religious, military, economic and agricultural elements. This sovereign, it seems, underwent ordeals with the gods themselves, made the crops grow, and imposed his supervision 'at every moment, over every person, every action, every thing' through the agency of the Mycenaean palace.[67] It may be questionable to discuss such a social order in terms of transcendence, but we may nonetheless wonder if the simulacra could possibly be an issue in such a thoroughly ruled and 'overcoded' society where villager and warrior-aristocrat never even come face to face.[68] There is no *isonomia* here, there is no equality, no democracy, and thus no problem of rivalry. What use could such a social order have for a concept of the counterfeit? Where is the need to distinguish the true rival from the false claimant? Everything from crops to war, and every person from farmer to warrior, is coded or determined by the *wanax* – so where, in the Mycenaean palaces of Knossos, Phaestos or Mallia, could the false, the counterfeit and the simulacral emerge? The simulacrum can manifest only within a social order where transcendence seems like a necessity – where phenomena are not coded *in advance*: 'The claimant calls for a ground; the claim must be grounded (or denounced as groundless). Laying claim is not one phenomenon among others, but *the very nature of every phenomenon*.'[69] The simulacra do not merely attest to a problem of the false and the counterfeit, they also imply that *the true has become a problem* – they imply that phenomena themselves stand in need of discernment. One needs the social space and time of an Athens, a 'crisis of sovereignty' where the *arche* required delegation

Difference and Representation

every year by free Athenian citizens (that is, citizens after Cleisthenes' democratic reforms), in order for the simulacrum to emerge.[70] In this very concrete sense, we may conclude that the simulacrum, acknowledged but subordinated to the copy by the Idea, can only emerge in an immanent field – it is a problem of democracy. The simulacrum cannot even appear in a simply transcendent order, and so the model-copy structure alongside which the simulacrum appears, far from being a transcendent structure, must preclude that very transcendence if the simulacrum is to remain visible.

The foregoing should be sufficient indication that, if Platonic Ideas introduce transcendence into immanence, this cannot be taken as the erection of a transcendent world over and above an immanent one. When Plato discovers the simulacrum 'in the flash of an instant', it is in his attempted solution to the immanent problem of rivalry in Athens, not because of ontological considerations of heavenly transcendence. Deleuze prefers the authenticating power of the Idea in Plato over the representational function of the categories in Aristotle, but this power of authentication is afforded through the immediate relation of difference (as opposed to the mediation of the categories). And if the Idea facilitates an immediate relation of difference, it is because it does not hold an external relation to different things – the difference between one thing and another is not grounded in a resemblance between them (as in Aristotle) but rather proceeds from the Idea directly to the thing – so the Idea, as internal to the relation of resemblance, cannot be transcendent to the things. We may profitably recall Lautman's Platonism here, not only because of his crucial influence on Deleuze's Idea-Problem complex but also because he rejected the two-world transcendent rendition of Platonic Ideas: 'Ideas are not immobile and irreducible essences of an intelligible world.'[71]

Lautman searched for a deeper problematic underneath the structure of model and copy within Platonic division, and instead of taking Plato's Ideas as 'models whose mathematical entities would merely be copies', he conceived of Ideas 'in the true Platonic sense of the term, [as] the structural schemas according to which the effective theories are organised'.[72] Lautman attempted to draw a portrait of mathematics as a realisation or solution of dialectical oppositions (or what he called 'dialectical notions') such as whole/part, continuous/discontinuous, global/local. So in the *Timaeus*, for example, the materials of which the universe is constituted are not elementary substances such as atoms but rather mixtures of Being and Becoming, the Same and the Different. These dialectical opposites are mixed according to harmonic and arithmetic means, ultimately constituting the image of the world.[73] Lautman thus observes that 'the properties of place and matter, according to him [Plato], are not purely sensible', but are the 'geometric and physical transposition of a dialectical theory'.[74] Space is a result of relations between dialectical notions, and Ideas are precisely the relations between dialectical notions. Hence, dialectical notions are dynamically composed – they are related or schematised differently according to the Ideas that bring the notions into composition. For Lautman, the priority of a dialectical composition is what ensures the intelligibility of the mathematical and the physical fields within which the dialectical relations are realised. In this fashion, Ideas compose a problem that is incarnated within effective mathematical or physical fields. Deleuze expresses his admiration for Lautman on this point, namely, the fact that Lautman conceived of problems as 'Platonic Ideas or ideal liaisons between dialectical notions, relative to "eventual situations of the existent"; but also how they are realised within the real relations constitutive of the desired solution within a *mathematical, physical* or other field'.[75] Mathematical theories are effective whenever they recall or 'recollect' the unity of the Ideational composition. On the other hand, to 'forget' the Ideational composition from which mathematics proceeds is to embark upon false dialectics – it is to 'forget' that 'the reality inherent to mathematical theories comes to them from their participation in an ideal reality that is dominating with respect to mathematics, but that is only knowable through it'.[76] Deleuze writes:

> Whenever the dialectic 'forgets' its intimate relation with Ideas in the form of problems, whenever it is content to trace problems from propositions, it loses its true power and falls under the sway of the power of the negative, necessarily substituting for the ideal objectivity of the problematic a simple confrontation between opposing, contrary or contradictory, propositions.[77]

Thus, the aforementioned 'third-man' criticism of the Idea was something of a 'forgetting' of the Idea in order to pursue a propositional method of division in terms of genera and species. The same may be said of the bifurcation of nature, which evaluates heterogeneous differences in time and materials as if they were all a question of homogeneous and superficial distinctions of shape, size, motion, space, etc. The basic problem lies in understanding difference through identity. In Aristotle, difference is managed through the identity of the general

Difference and Representation

concept or the identity of resemblance between two different things, so the resemblance between Meno and Anytus is conceived in terms of the species, man. For Plato, on the other hand, the question is to discern the Ideational composition that Meno participates in, like a son participates in the Ideational composition of his father. It behooves us to remember that the most common words 'Idea' translates, *eidos* and *idea*, have ocular and visual significance; they could be understood as the 'look' of something. To find out who Meno is would be to recollect through the dialectic the Ideas that compose Meno. It is to discern a principle of filiation that lies behind or dominates the manifestations of Meno, the 'look' of Meno that his appearances 'participate' in or 'come after', appearances in both word and deed.

In a well-known talk called 'The Method of Dramatization', Deleuze argued that the question, '*What is this?*', is not a good question for discovering the Idea, and that it 'reveals itself to be confused and doubtful, even in Plato and the Platonic tradition. Because the question *What is this?* is in the end the driving force behind those dialogues known as aporetic.'[78] In fact, Deleuze continues, 'when the Platonic dialectic becomes something serious and positive, it takes other forms: who? in the *Republic*; how much? in the *Philebus*; where and when? in the *Sophist*; and in which case? in *Parmenides*'.[79] For Deleuze, of course, the idea 'is much closer to the accident than to the abstract essence, and can be determined only with the questions *who? how? how much? where and when? in which case?* – forms that sketch genuine spatio-temporal coordinates of the Idea', that is to say, the concrete situation of the Idea.[80] But even in Plato, Ideas do not occupy a transcendent location in some ideal realm, as essences reigning above appearances. The peculiar transcendence of the Platonic Idea, functioning within the immanent field of the agora, lies elsewhere – specifically, in its mode of evaluation. For Plato's Ideas evaluate rival claims in terms of an original that has the form of an identity (denoted as *auto kath' hauto*, 'itself by itself'). In this regard, Deleuze finds that the Ideas are traced upon the outlines of the empirical, as if the Form of something maintained the identity attributed to the sensible thing itself. This can be examined through the distinction between two types of memory in Plato. In the *Meno*, for example, the handsome, rich and well-born Meno also has a good memory for words – he remembers, for example, what Gorgias has said. Socrates, on the other hand, has forgotten what Gorgias has said.[81] While Meno remembers words, the dressings of meanings,

Socrates forgets words, but recollects Ideas. But far from being a deficiency, as John Sallis observes, Socrates' forgetting of words has a positive function because he forgets words in order to recollect what was spoken about in the words.[82] The distinction here is one between empirical memory, which facilitates *recognition* but does not disturb thought, and recollection, which is the exercise provoked by an encounter that forces us to think. Empirical memory deals with things that must have been sensed or experienced in some fashion. Meno remains ignorant of his own ignorance because he is content to uncritically repeat what he has heard or seen in the past – precisely the function of recognition – 'this is a finger, this is a table, Good morning Theaetetus'.[83] What Socrates recollects, on the other hand, has not occurred in the empirical past. As Deleuze notes, recollection in Plato 'grasps that which from the outset can only be recalled, even the first time: not a contingent past, but the being of the past as such and the past of every time'.[84] In this sense, there is something in the sensible itself that points beyond the empirical, not toward a sensible thing but rather the very condition of the sensible:

> [The object of the recollection] is not a sensible being but the being *of* the sensible. It is not the given but that by which the given is given. It is therefore in a certain sense the imperceptible [*insensible*]. It is imperceptible precisely from the point of view of recognition – in other words, from the point of view of an empirical exercise of the senses in which sensibility grasps only that which also could be grasped by other faculties...[85]

In other words, the empirical thing – a finger, a table, Theaetetus – is something amenable to all the faculties: we can think of them, we can sense them, we can remember them. But Deleuze searches for the being of the sensible as that which can only be sensed, in fact, this is the objective problem of *transcendental* empiricism. But Deleuze constantly reminds us that we cannot trace the transcendental on the outlines of the empirical. We finally arrive at Deleuze's criticism of transcendence in the Platonic Idea – Plato imputed to recollection an object in the mould of an empirical original. In other words, the original recollected instance is like a past being, the only difference is that Plato substitutes a *mythological* past for an empirical one. Consequently, Deleuze finds that 'reminiscence [recollection] is still a refuge for the recognition model, and Plato no less than Kant traces the operation of the transcendental memory from the outlines of its empirical exercise'.[86]

Deleuze's reversal of Platonism thus recovers essential dimensions

Difference and Representation

of Platonism, such as the immediate relation of difference and the immanence of the Idea. Ideas in Plato served an evaluative function, not a representational one. For Deleuze, 'Ideas are not simple essences, but multiplicities or complexes of relations.'[87] And he explicitly recalls the Riemannian and Bergsonian usage of the term:

> Ideas are multiplicities: every idea is a multiplicity or a variety. In this Riemannian usage of the word 'multiplicity' (taken up by Husserl, and again by Bergson) the utmost importance must be attached to the substantive form: multiplicity must not designate a combination of the many and the one, but rather an organisation belonging to the many as such, which has no need whatsoever of unity in order to form a system ... Everywhere the differences between multiplicities and the differences within multiplicities replace schematic and crude oppositions.[88]

One such crude opposition that must be overcome is the distinction between mind and matter, between the subject and the object, between the human and the non-human. The concept of difference, which is the internal working of Ideas/multiplicities, dismantles the brute dichotomies that rely on identity, the framework of representation. Instead of differences between things, we have only differences that produce differences. But how, then, are things and identities produced? In the next chapter, we examine the concepts of technology and events as a way of accounting for the production of subjects, objects and the other identities that we abstract from the concrete process of becoming. Nonetheless, things, in the representational sense of the word, are merely abstractions from differential processes, a process of becoming characterised by Ideas. And just as Marx said that we cannot discern the process of wheat production simply from the taste of the wheat, so too we cannot trace the process of production on the outlines of the products produced. The traditional dichotomies between the subject and object, the human and the non-human, mind and matter, and the natural and the artificial, thus stand in need of reevaluation. Such a reevaluation, if positive, would underscore the fact that those dichotomies, along with the identities they isolate as their terms, are technical products of a process that subsumes them within a multiplicity of differential relations. So underneath these dichotomies we find a wilder nature of Ideas, and underneath techniques of representation and axiomatisation, a problematic field of differences that operates in a sub-representational domain.

Notes

1. Deleuze, *Difference and Repetition*, p. 69.
2. Deleuze, *Bergsonism*, p. 27. (Bergson quoted from *Matter and Memory*, p. 184). Regarding his own system of Ideas, Deleuze writes: 'Here, too, we follow the path to the bend at which "reason" plunges into the beyond.' Deleuze, *Difference and Repetition*, p. 282.
3. Henri Bergson, *The Creative Mind: An Introduction to Metaphysics*, trans. M. Andison (Mineola: Dover Publications, 2007), p. 81. Cf. Bergson, *Creative Evolution*, pp. 272–98.
4. Daniel Smith, 'The Concept of the Simulacrum: Deleuze and the Overturning of Platonism', *Continental Philosophy Review*, Vol. 38 (2006), p. 105.
5. De Beistegui provides an excellent examination of the concept of the simulacrum found in the Stoics and Lucretius, and he presents Deleuze's reversal of Platonism in terms of Lucretian metaphysics: 'By saying that identity is a product of difference, Deleuze not only reverses the terms of Platonism. He also, and most significantly, replaces the metaphysics of representation with a metaphysics of production, which Lucretius had already intimated: difference, he claims, is essentially productive, and the only engine or principle of production.' Miguel de Beistegui, 'The Deleuzian Reversal of Platonism', in *The Cambridge Companion to Deleuze*, ed. Daniel Smith and Henry Somers-Hall (Cambridge: Cambridge University Press, 2012), p. 73.
6. Deleuze, *Difference and Repetition*, p. 29.
7. Aristotle, Metaphysics Γ.2, in *The Complete Works of Aristotle, Vol. 2*, ed. J. Barnes (Princeton: Princeton University Press, 1984), p. 1584.
8. Deleuze, *Difference and Repetition*, p. 34.
9. For exegetical and constructive accounts of individuation in Aristotle, see T. Scaltsas, D. Charles, and M. L. Gill, *Unity, Identity, and Explanation in Aristotle's Metaphysics* (Oxford: Clarendon Press, 1994). See especially Aryeh Kosman's 'The Activity of Being in Aristotle's *Metaphysics*', which develops a compelling genetic account of unity and individuation (through the concepts of *dunamis* [potentiality] and *energeia* [actuality]) that would resist certain points of Deleuze's interpretation of Aristotelian individuation in *Difference and Repetition*.
10. Daniel Smith, 'The Doctrine of Univocity: Deleuze's Ontology of Immanence', in Mary Bryden (ed.), *Deleuze and Religion* (London: Routledge, 2001), p. 177. Deleuze makes this point in *Difference and Repetition*: 'it (analogy) retains in the particular only that which conforms to the general (matter and form), and seeks the principle of individuation in this or that element of the constituted individual' (p. 38).

Difference and Representation

11. Aristotle, *The Complete Works of Aristotle, Vols. 1 & 2*, ed. J. Barnes (Princeton: Princeton University Press, 1984). Italics mine. All citations of Aristotle are from this edition.
12. See Umberto Eco, *Kant and the Platypus*, trans. Alastair McEwen (San Diego: Harcourt, Inc., 1999). Eco tells us that Aristotle would have been in a compromising position with regard to the Platypus, 'because, since he would have been convinced that a platypus had to have an essence independent of our intellect, the impossibility of finding a definition for it would have disquieted him all the more' (p. 90).
13. Deleuze, *Difference and Repetition*, p. 59.
14. Deleuze, *Difference and Repetition*, pp. 59–60; italics mine.
15. On the complicated notion of participation in Plato, see Allan Silverman, *The Dialectic of Essence: A Study of Plato's Metaphysics* (Princeton: Princeton University Press, 2002), and Fritz-Gregor Herrmann, *Words and Ideas: The Roots of Plato's Philosophy* (Swansea: The Classical Press of Wales, 2007).
16. Deleuze, *Difference and Repetition*, p. 60.
17. Deleuze, *Difference and Repetition*, p. 60.
18. Deleuze, *Difference and Repetition*, p. 61.
19. Deleuze, *Difference and Repetition*, p. 61.
20. Deleuze, *Difference and Repetition*, p. 61.
21. Each of the four arguments or discourses of the *Phaedo* are preceded by a story-telling: the discourse from opposites by the fabled Aesopian story of pleasure and pain; the discourse from recollection by the myth of metempsychosis; and the discourse from invisibility by the story of children's songs. Interestingly, for the fourth discourse from causes, after pausing for a period of time, Socrates narrates his intellectual autobiography.
22. Gilles Deleuze, *The Logic of Sense*, trans. M. Lester and C. Stivale (New York: Columbia University Press, 1990), p. 256; italics mine.
23. Friedrich Nietzsche, *Twilight of the Idols and The Anti-Christ*, trans. R. J. Hollingdale (Middlesex: Penguin, 1968), p. 41.
24. Deleuze, *Logic of Sense*, p. 253.
25. Deleuze, *Difference and Repetition*, p. 59.
26. Deleuze, *Difference and Repetition*, p. 59.
27. Plato explores the question of the fitness and truth of language in the *Cratylus*, where Socrates notes that language is imperfect because words or names are not what they signify, that is to say, they neither coincide with things nor with the forms. 'And you also agree that a name is an imitation (*mimema*) of a thing?' *Cratylus*, 430b in Plato, *Complete Works*, ed. J. Cooper (Indianapolis: Hackett Publishing Company, 1997). In fact, words can be deceptive precisely because they are neither forms nor things (appearances), but rather imitations. It would perhaps be interesting to examine the *Cratylus* vis-à-vis Nietzsche's overturning

of Platonism, since words imitate things that supposedly are, so it seems to be the case that the difference or space between words and things is what affords the adventure of language in the first place. Consequently, imitation is not the ground for truth, equation, correspondence or adequacy to the forms, but rather for the unfolding of the *logos* itself. Cf. *Republic*, 382ff. On the space between names and things, and a penetrating discussion of *logos* in the *Cratylus*, see John Sallis, *Being and Logos: Reading the Platonic Dialogues*, 3rd edn (Bloomington: Indiana University Press, 1996), pp. 278–9/183–305.

28. Friedrich Nietzsche, 'On Truth and Lies in a Nonmoral Sense' (1873), in *The Nietzsche Reader*, ed. Keith Ansell Pearson and Duncan Large (Oxford: Blackwell Publishing, 2006); 'den Worten nie auf die Wahrheit', continues with 'nie auf einen adäquaten Ausdruck ankommt'. Thus, the statement announces its opposition to Aquinas' 'Veritas est adaequatio intellectus et rei' as well as Kant's 'Ding an sich' referred to in the following line (Das 'Ding an sich' (das würde eben die reine folgenlose Wahrheit sein) ist auch dem Sprachbildner ganz unfasslich und ganz und gar nicht erstrebenswerth.). However, while the references here seems to focus on Kant and the medieval notion of adequation, the general trajectory of Nietzsche's essay clearly isolates Platonism as its ultimate target.
29. Nietzsche, *Twilight of the Idols*, p. 37.
30. Nietzsche, 'On Truth and Lies', p. 116.
31. Nietzsche, 'On Truth and Lies', p. 117; italics mine.
32. Nietzsche, 'On Truth and Lies', p. 117.
33. While Plato uses '*metechein*' in *Laches*, *Charmides*, *Protagoras* and *Gorgias*, the notion of participation is complicated by the use of '*pareinai*' and '*paragignesthai*' (*Gorgias*, *Charmides*), and, in the *Phaedo*, '*koinōnia*'.
34. Whitehead, *The Concept of Nature*, p. 25.
35. Whitehead, *Science and the Modern World*, pp. 50–1.
36. Here again we find remarkable connections with the *Cratylus*, where Hermogenes states that names arise from convention and agreement, and ultimately by law (*nomos*) and custom (*ethos*). Cf. Sallis, *Being and Logos*, pp. 192–5.
37. Nietzsche, 'On Truth and Lies', p. 115.
38. Nietzsche, 'On Truth and Lies', p. 118.
39. Nietzsche, 'On Truth and Lies', p. 119.
40. Nietzsche, 'Homer's Contest', in *The Nietzsche Reader*, p. 97.
41. Jean-Pierre Vernant, *The Origins of Greek Thought* (Ithaca: Cornell University Press, 1982).
42. Vernant, *The Origins of Greek Thought*, p. 50. See especially chapter 4, 'The Spiritual Universe of the *Polis*', where he identifies speech as the primary tool that facilitated political contests in the Greek *polis*. See also Smith, 'The Concept of the Simulacrum'.

43. Vernant, *The Origins of Greek Thought*, p. 101.
44. We will briefly discuss the concept of technology as it relates to architecture in the following chapter.
45. Pierre Lévêque, *Cleisthenes the Athenian: An Essay on the Representation of Space and Time in Greek Political Thought from the End of the Sixth Century to the Death of Plato* (Amherst: Humanity Books, 1997), pp. 9–22.
46. Lévêque, *Cleisthenes the Athenian*, p. 83.
47. Pierre Vidal-Naquet, *The Black Hunter: Forms of Thought and Forms of Society in the Greek World*, trans. Andrew Szegedy-Maszak (Baltimore: The Johns Hopkins University Press, 1986), p. 257.
48. Vernant, *The Origins of Greek Thought*, p. 47.
49. Nietzsche, *Twilight of the Idols*, p. 40.
50. Deleuze, *Difference and Repetition*, p. 127; italics mine.
51. Deleuze, *Logic of Sense*, p. 257. Cf. Smith, 'The Concept of the Simulacrum', p. 99.
52. Deleuze, *Logic of Sense*, p. 256.
53. To be sure, the Idea in Deleuze is constructed not merely through Plato, but also through Kant, Lautman and others. However, the Platonic inheritance of the Idea in Deleuze is largely dismissed. Part of what I would like to argue here is that attending to the Platonic inheritance of the Idea in Deleuze opens up productive lines of inquiry into both Deleuze and Plato.
54. Deleuze, *Logic of Sense*, p. 253.
55. Deleuze, *Difference and Repetition*, p. 59; italics mine.
56. Deleuze, *Logic of Sense*, p. 258.
57. Deleuze, *Difference and Repetition*, p. 17.
58. Gilles Deleuze, *Nietzsche and Philosophy* (New York: Columbia University Press, 2006), p. 47.
59. Deleuze, *Difference and Repetition*, p. 277.
60. Deleuze, *Difference and Repetition*, p. 128.
61. Deleuze, *Logic of Sense*, p. 256.
62. Deleuze, *Difference and Repetition*, p. 128.
63. Sean Bowden dedicates a chapter of his study of *The Logic of Sense* to Lautman and Simondon's influence on Deleuze's Problem-Idea-Singularity complex. See Sean Bowden, *The Priority of Events: Deleuze's Logic of Sense* (Edinburgh: Edinburgh University Press, 2011), especially pp. 110–17. See also Daniel Smith's excellent article on the Kantian inheritance of Deleuze's Idea in 'Deleuze, Kant, and the Theory of Immanent Ideas', in *Essays on Deleuze* (Edinburgh: Edinburgh University Press, 2012), pp. 106–21.
64. Gilles Deleuze, *Essays Critical and Clinical*, trans. Daniel Smith and Michael Greco (Minneapolis: University of Minnesota Press, 1997), p. 137.

65. Deleuze, *Essays Critical and Clinical*, p. 137.
66. M. I. Finley, 'Homer and Mycenae: Property and Tenure', *Historia*, Vol. 6 (1957), pp. 133–59.
67. Vernant, *The Origins of Greek Thought*, p. 41.
68. Vidal-Naquet, *The Black Hunter*, p. 255.
69. Deleuze, *Difference and Repetition*, p. 62; italics mine.
70. On the crisis of sovereignty see Vernant, *The Origins of Greek Thought*, chapter 3.
71. Lautman is following the work of Robin, Stenzel and Becker here with regard to the role of numbers and Ideas. Albert Lautman, 'Essay on the Notions of Structure and Existence in Mathematics', in *Mathematics, Ideas and the Physical Real*, trans. Simon Duffy (New York: Continuum, 2011), p. 190.
72. Lautman, *Mathematics, Ideas and the Physical Real*, p. 199.
73. See F. M. Cornford, *Plato's Cosmology* (London: Routledge & Kegan Paul Limited, 1948); and John Sallis, *Chorology: On Beginning in Plato's Timaeus* (Bloomington: Indiana University Press, 1999). For Deleuze on the *Timaeus*, see *Difference and Repetition*, p. 233.
74. Lautman, *Mathematics, Ideas and the Physical Real*, p. 231.
75. Deleuze, *Difference and Repetition*, pp. 163–4.
76. Lautman, *Mathematics, Ideas and the Physical Real*, p. 30.
77. Deleuze, *Difference and Repetition*, p. 164.
78. Gilles Deleuze, 'The Method of Dramatization', in *Desert Islands and Other Texts, 1953–1974* (Los Angeles: Semiotext(e), 2004), p. 95.
79. Deleuze, 'The Method of Dramatization', p. 95.
80. Deleuze, 'The Method of Dramatization', p. 96.
81. Plato, *Meno*, 71c-d.
82. Sallis, *Being and Logos*, p. 68. Sallis also notes that Meno's name is itself a derangement of the word for memory, *mneme*.
83. Deleuze, *Difference and Repetition*, p. 138.
84. Deleuze, *Difference and Repetition*, p. 140.
85. Deleuze, *Difference and Repetition*, p. 140.
86. Deleuze, *Difference and Repetition*, p. 142.
87. Deleuze, *Difference and Repetition*, p. 163.
88. Deleuze, *Difference and Repetition*, p. 182.

4

Beyond the Nature-Artifice Divide: Technology, Milieu and Machine

> 'My poor child, do you want me to tell you the truth?
> I've been given a name that does not suit me:
> For I am called Nature, yet I am all art.'
> Voltaire, *Philosophical Dictionary*

In a 1988 interview, Deleuze indicated that he would like to return to joint work with Guattari 'and produce a sort of philosophy of Nature, now that any distinction between nature and artifice is becoming blurred'.[1] My intention here is not to ask what such a philosophy of nature might look like, but only to introduce a few concepts that would be helpful for dismantling the division between nature and artifice. This division, alongside the ancillary distinctions between mind and matter, subject and object, is the hallmark of representational philosophies of nature. Such divisions facilitate the disciplinary classification of knowledge in general, typically cast in terms of a hierarchy of sciences that ascends from physics, chemistry and biology (the natural sciences) to anthropology, sociology and economics (the social sciences). But humans must first be separated from nature in order to have sciences of humans over and against sciences of nature.[2] Moreover, while each of these sciences enjoys different methods and objects of inquiry (whether physical, biological or economic), the hierarchical arrangement of the sciences considers nature as a substantial foundation upon which human artifice operates and to which it responds. By focusing on events rather than substances, the concepts of technology, milieu and machine suggest a different way of considering the complicated relation between the various registers of knowledge, beyond the hierarchical classification that turns upon the division between the natural and the artificial.[3] Furthermore, these event-based concepts should prove useful when drawing the profile of a philosophy of nature that is not only more plausible than representational alternatives, but also more ecologically responsible as well.

Technology

THE CONCEPT OF TECHNOLOGY IN GENERAL

In an 1982 interview with Paul Rabinow on architecture, Foucault laments the fact that our use of the word 'technology' is confined to such a narrow meaning of 'hard technology, the technology of wood, of fire, of electricity', but considers that if we disabuse the term of its narrow confines, we find that '*government* is also a function of technology: the government of individuals, the government of souls, the government of the self by the self...'[4] To be sure, there seems to be a prima facie distinction to be made between the material logistics of 'hard technologies', of instruments and machines, on the one hand, and the social institutions of government (despite its seemingly mechanical and insensitive bureaucratic procedures), on the other. But Foucault endorses a wide concept of technology, *techne* understood generally as 'a practical rationality governed by a conscious goal'.[5] If we understand a material instrument as an artifact constructed for a certain goal, as a clock is built to tell the time, and if we understand government as an artifice constructed to conduct human activity, then the technological dimension of both the material instrument and the social institution comes into view: both the clock and government are practical and artificial constructions that rationally conduct or govern us toward a conscious goal. The concept of technology thus enjoys the critical capacity to bypass the boundary between the social and the material, the human and the non-human, which in turn allows us to conceive a world order of things unencumbered by the bifurcation between the natural and the artificial.

In fact, Foucault's general concept of technology recalls the Aristotelian formulation of *techne*, which often holds material elements alongside physiological and political elements. Consider the virtue of *megalopsychia*, or 'greatness of soul', surely one of the most striking (and least discussed) virtues in the *Nicomachean Ethics* (4.3. 1123b–1125a).[6] Aristotle is tellingly detailed when describing the *megalopsychon*, or the great-souled person: 'a slow step is thought proper to the proud man, a deep voice, and a level utterance; for the man who takes few things seriously is not likely to be hurried, nor the man who thinks nothing great to be excited, while a shrill voice and a rapid gait are the results of hurry and excitement' (*NE* 1025a12–16). While the vain person (*chaunotes*) indulges in honours from 'casual people and on trifling grounds', and while the humble person (*mikropsychion*) refuses gifts and honours that he nonetheless deserves, the *megalopsychon* accepts great gifts and honours in a great way, at a great time, etc. We are presented here with an art of giving and receiving honourably, a style or *techne* of character composition that seems to be analysable in terms of gait, step, voice, prosody, breathing and speed.[7]

To be sure, I do not mean to delve into the finer complexities of the *Nicomachean Ethics* here; rather, I mention the walking *megalopsychon* because it can guide us through a wider concept of *techne* that Foucault endorses, *techne* beyond the opposition between the natural and the artificial. Aristotle's evaluation of the *megalopsychon* can illuminate the way the concept of technology is critical of the natural/artificial distinction precisely because the sociopolitical phenomenon of becoming *megalopsychon* involves *time and materials*: it is constituted by a certain technology of walking slowly through ancient Greece, perhaps with an appropriate train, and it requires the artful reception of politically valorised material objects, that is to say, gifts, with a technology of the body – a deep voice and a level prosody. Of course, it is not just the *megalopsychon* in the agora who engages walking as a techno-political activity, we too move in a social space, dressed in some sort of attire, outfitted with shoes and canes and wheelchairs. Furthermore, alongside these technological prosthetics, we have a certain *way* of carrying ourselves which cultivates a certain affect, in the sense that we are able to affectively respond to the ways that other people acknowledge or greet us on our walks, and we have practical techniques for avoiding physical accidents as well as social blunders. Both these material and institutional artifices form a concrete ensemble of technologies that produce the *event* of walking down the street through town.

And it is in this regard that Foucault suspects that 'if one placed the history of architecture back in this general history of *techne*, in this wide sense of the word, one would have a more interesting guiding concept than by the opposition between the exact sciences and the inexact ones'.[8] In other words, Foucault reckons that the concept of technology (taken in the general sense) would allow us to examine the relation between humans and the non-human material architectures surrounding them in a better fashion than by way of the confrontational opposition between the hard sciences and the social sciences and humanities. But Foucault's suspicion is not only a suggestion for the discipline of architectural history, it is also an invitation to reevaluate divisions between the human and the non-human, the division between social artifices and natural substances. Later in this chapter we will examine this in more detail but, for now, we note that this wide sense of *techne* shares a critical capacity with Deleuze and Guattari's concept of machine in *Anti-Oedipus*, since machines also dismantle the traditional bifurcation of nature and society.

In his case history, 'Joey: A "Mechanical Boy"', Bruno Bettelheim relates the story of a schizophrenic child whose performance of 'intrinsically human' actions 'never appeared to be other than machine-started and executed'.[9] While Bettelheim focuses on the abnormal emotional development of a child ignored by his mother in a mechanised society, Deleuze and Guattari examine Bettelheim's case study as a testimony of the *machinic* nature of desire itself: little Joey can eat, sleep and communicate only by being plugged into machines. In other words, the events that constitute the state of affairs surrounding little Joey are not the expression of some internally hidden unconscious psyche but rather the productions of a process conducted, quite literally, by machines. Ultimately, Deleuze and Guattari's concept of machine allows them to conclude that 'man and nature are not like two opposite terms confronting each other . . . rather, they are one and the same essential reality'.[10]

Uncertainty and the Management of Time and Society

In *Security, Territory, Population*, Foucault undertakes a brief examination of the capitalisation of towns, beginning with a case study of Pierre Lelievre's planning of Nantes in 1932. There, the question of technology is broached within the context of Foucault's analysis of sovereignty, discipline and security. While '*sovereignty* capitalises a territory, raising the major problem of the seat of government', and while '*discipline* structures a space and addresses the essential problem of a hierarchical and functional distribution of elements', '*security* will try to plan a milieu in terms of *events or series of events or possible elements* . . . the specific space of security refers then to a series of possible events, it refers to the temporal and the uncertain which have to be inserted within a given space . . . the milieu'.[11] Security thus addresses the problem of how to manage a series of events within a civic space, forming a kind of diagram for the town that considers time and its various possibilities. Foucault observes that the development of Nantes, which unfolds according

Beyond the Nature-Artifice Divide

to an idea of security, an idea of planning and preparing for the possible events that may occur in the future, constitutes a technical problem, again in the wider sense of *techne* because the town must be organised in a practical way oriented toward a conscious goal. The problem of developing Nantes, for example, took into consideration *hygiene* (since pockets of dense population had to be opened up to ventilation); *trade* (since the streets had to facilitate the transportation of goods into the town and outside of the town); *traffic* (since the streets had to open onto outside roads that had to be manageable in order for customs to maintain control of trade); and *surveillance* (since the suppression of city walls made necessary by economic development entailed that one could no longer close towns in the evening or closely supervise daily comings and goings, which meant that criminals might creep in at any time from the country – which is of course, Foucault quips, where criminals come from).[12] The essential point here is that in each of these cases, the technical problem of town development is oriented toward time and knowledge: 'What must be done to meet something that is not exactly *known in advance*?'[13] Technical problems are problems that concern the eventual as the *limit of knowledge*, because the eventual future, when apprehended in terms of knowledge, harbours an element of uncertainty. From an epistemic perspective, in other words, the future signifies uncertainty, an uncertainty that needs to be managed and conducted – and we find that *techne* is precisely the way to conduct the present into the future. In this regard, *techne* is a mode of intervening upon becoming, it is a way of ordering time and events.

Technology, Knowledge and Time

In the *Nicomachean Ethics*, Aristotle distinguishes *techne* from *episteme* (knowledge derived from first principles) in terms of variability: *techne*, or art concerned with making, treats things that admit of variation, while *episteme*, or knowledge that can be demonstrated (e.g. geometrical proofs), treats things that are invariable (*NE* 6.4–6. 1140a–1141a). In other words, *episteme* is concerned with 'things that are universal and necessary' (*NE* 6.6. 140b30), while *techne*, as Heidegger observes, 'is concerned with beings only insofar as they are in the process of becoming' (*NE* 6.4. 1140a10–11; cf. *Metaphysics* Z.7. 1032a12–13).[14] Thus, the distinction between *techne* and *episteme* in terms of variability, contingency and necessity harbours a more profound distinction in terms of time: *techne* deals with things as they are coming into being while *episteme* deals with invariant truths of things that are.

But this is all to say that from an epistemological perspective, the process of becoming (as contingency) presents *episteme* with a limit, a lack of knowledge, an element of uncertainty. From this epistemic perspective, *techne* can be seen as an attempt to remedy or compensate for this limitation – we technologically conduct events to know what otherwise cannot be known in advance. But from a temporal and metaphysical perspective, we may understand contingency (and thus uncertainty) in a productive register. On this account, contingency is essentially the novelty of becoming, the fact that the new cannot be wholly determined by the known. Contingency is thus understood as a creative eruption at the edges of knowledge; in other words, contingency and uncertainty are not mere *deficiencies* of knowledge but rather *productive effects* of the essentially creative advance of becoming. The distinction between the epistemic perspective and the temporal/metaphysical perspective reveals the socio-political dimension of knowledge, since the epistemic perspective treats contingent events as something to be governed and conducted through techniques. The rationality of techniques thus emerges from within a society invested in determining and conducting the future in advance, and so as we saw with Nietzsche in the last chapter, we discover the will to know as a socio-political will. *Techne*, in this sense of practical rationality, is a mode of intervening upon becoming within the context of a social order, a mode of conducting events in order to *determine* precisely those aspects of the future that are not *knowable* in advance. The 'rationality' of *techne* is sociopolitical rationality rooted in a set of practices aimed at conducting so many contingencies and becomings that lie beyond the purview of *episteme*, yet within the purview of security and determinability. And since the rationality of technology is rooted in the social order, particular technologies remain unintelligible apart from the social orders within which they hold their function of conducting events.

It is telling, then, that architecture is the context from which Foucault endorses *techne* as a better guiding concept than the distinction between exact and inexact sciences. Aristotle tells us that architecture is a *techne* because it is a rational quality concerned with making (*NE* 6.4 1140a). And while we can imagine how *episteme* might convene with *techne* in architecture, as geometric knowledge of cylinders and regular polygons may aid in the building of the

Beyond the Nature-Artifice Divide

Parthenon, Aristotle's aim is to distinguish *techne* from *episteme* by reference to architecture's concern with things that are in the process of becoming. But this reveals the fundamentally temporal dimension of architecture – it is not simply the art of the becoming or production of a spatial edifice, rather, it is more profoundly an art of the arrangement of time and events. What is at stake in the construction of the Parthenon is not, after all, simply the production of a building, but rather the determination of events, affects and significance – it is a temple, a symbol, a landmark, a place, in other words, built through a practical rationality oriented toward conducting a series of social affects and events. The point here is not to make the obvious claim that the arrangement of space is value-laden; the point is rather that Foucault's suggestion to reintroduce *techne* as a guiding concept for architecture exhibits the fundamentally temporal orientation of designing space: the arrangement of space is secondary to the determination of events and affects. Even when building a home, one is attempting to manufacture a sequence of affects and events called 'living at home' that would emanate from and converge upon the space of the house. This is what it means to say that *techne* is properly concerned with time and becoming; it is a mode of arranging events or intervening upon becoming.

Milieu

THE CONCEPT OF MILIEU IN GENERAL

Hippocrates, in his *On Airs, Waters, and Places*, encourages the student of medicine, upon arriving 'into a city to which he is a stranger', to 'consider its situation, how it lies as to the winds and the rising of the sun; for its influence is not the same whether it lies to the north or the south, to the rising or to the setting sun . . . From these things he must proceed to investigate everything else' (I–II).[15] What is striking is not merely the course of Hippocratic investigation, which starts by reading the geographical environment in order to decipher 'the diseases peculiar to the place, or the particular nature of common diseases'; but also its overarching resolve: 'if it shall be thought that these things belong rather to meteorology, it will be admitted, on second thoughts, that astronomy contributes not a little, but a very great deal, indeed, to medicine. For with the seasons the digestive organs of men undergo a change' (II), and it is through the medium of the digestive organs, Hippocrates contends, that climate influences the characters and dispositions of the various races of people (XXIII).[16] The fulcrum of the Hippocratic treatise, then, is to comprehend the changes in man through changes in the world, because the causes of human malady do not lie exclusively within the body, rather, they lie quite distantly outside of it, distributed throughout the surrounding geography and even, farther off, among the stars.[17]

Accordingly, Hippocrates provides a classification of races that brings together various *anthropological* elements, such as physiological and political constitutions, as well as various *geographical* elements surrounding cities, such as their position with regard to the sun and waters, as well as the seasonal dews and breezes that characterise them:

> The races in Europe differ from one another, both as to stature and shape, owing to the changes of the seasons . . . These changes are likely to have an effect upon generation in the coagulation of the semen, as this process cannot be the same in summer as in winter, nor in rainy as in dry weather; wherefore, I think, that the figures of Europeans differ more than those of Asiatics . . . and the same may be said of their dispositions, and therefore I think the inhabitants of Europe more courageous than those of Asia . . . where the changes of season are very great, [the inhabitants are] . . . naturally of an enterprising and warlike disposition.[18]

But how, exactly, are we to account for the interaction between such vast and heterogeneous elements like airs, waters, digestive organs, semen, wars, cities and stars? To be sure, in Hippocrates, the influence is one-way from the environment to the individual, but the problem motivating his treatise is clearly legible: how does change circulate throughout the various registers between organism and environment? In other words, how are heterogeneous series of events (the seasons, organic processes, campaigns of war) linked together?

As we saw above, the development of Nantes constituted a technical problem because a series of uncertain events had to be managed in advance. The town, that is, must be *planned*, which is to say that the future, along with the complex of heterogeneous elements that it involves, must be coordinated within the social space of the town. Foucault calls this 'space in which a series of uncertain elements unfold' a *milieu* – it is the spatial arrangement of rivers, streets and institutions (with regard to hygiene, trade, traffic and surveillance, in the case of Nantes) that attempts to orchestrate events and affects:

> The milieu is a set of natural givens – rivers, marshes, hills – and a set of artificial givens – an agglomeration of individuals, of houses, etcetera . . .

Beyond the Nature-Artifice Divide

> What one tries to reach through this milieu, is precisely the conjunction of a series of events produced by these individuals, populations, and groups, and quasi natural events which occur around them.[19]

The milieu, then, is the site of the conjunction between the so-called 'natural givens' and 'artificial givens'. The character of the events that occur within the space of this conjunction is therefore heterogeneous, or 'quasi-natural'. While *techne* is a practical rationality that attempts to intervene upon eventualities, the *milieu* is the field of determination for these eventualities – the space, as it were, of technological intervention. We may say, then, that the milieu (literally, the 'middle place') is a live space that is composed by the heterogeneous series of organism and environment through the artifices that conduct their mutual eventualities. The milieu is thus not so much a given space, but a space composed through the conjunction of heterogeneous eventualities.

Of course, the concept of the milieu is not native to urban planning. In this section, I provide a short history of the concept of milieu, from seventeenth-century natural philosophy up to its reformulation by Deleuze and Guattari in terms of the concept of machine. As we shall see, the concept of milieu was introduced into biology through Lamarck and Comte but, as Foucault reminds us, it was already operative in Newtonian mechanics.

ACTION AT A DISTANCE: FROM DESCARTES TO NEWTON

The milieu, Foucault tells us, is 'what is needed to account for action at a distance of one body on another. It is therefore the medium of an action and the element in which it circulates.'[20] The concept of action at a distance is one of the most controversial issues in the history of mechanics, because it is not at all clear how distinct physical bodies can affect each other across distances, as is apparently the case with phenomena such as gravitational or magnetic attraction.

In Query 31 of the *Opticks*, Newton asks, 'Have not the small particles of bodies certain powers, virtues or forces, by which they act at a distance'?[21] For Newton, apparent action at a distance was something that required experiment before explanation, and so the empirical testimony of gravitational, magnetic and electric phenomena granted at least the *possibility* of action at a distance, despite the impossibility that natural philosophers encountered when trying to imagine it. Continental natural philosophers, most famously

Leibniz, understandably criticised Newton for reintroducing occult qualities that threatened the hard-won rational picture of the world – for how can action at a distance even be conceived? If we perceive two distinct bodies, such as a pair of magnets, acting at a distance, is there not a material explanation that connects them? Does not reason fill in the gap left by the senses? Championing the principle of action by physical contact, Leibniz maintained that while empirical phenomena such as magnetic attraction seem to display action at a distance, such phenomena must be rationally interrogated until they confess motion by contact: 'a body is never moved naturally, except by another body which touches it and pushes it; after that it continues until it is prevented by another body which touches it. Any other kind of operation on bodies is either miraculous or imaginary.'[22] Newton's defender Samuel Clarke did not merely agree with Leibniz, he took the principle of contact to betray the rule of logic: 'That one body should attract another without any intermediate means, is indeed not a miracle, but a *contradiction*: for 'tis supposing something to act where it is not.'[23] The dispute, of course, did not regard the apparent fact of action at a distance, since the appearance of action at a distance is perfectly clear. Rather, the dispute revolved around the question of knowledge, of what could count as knowledge. If the world is material, then it follows the principles of mechanism, and so knowledge of the world must be mechanistic. The concept of action at a distance seemed to go against the conditions of mechanistic knowledge; consequently, it was not only inconceivable, but also a considerable threat to the image of rationality. Corpuscular philosophy and Newton's empiricism were not equal rivals, because the corpuscular system was the standard of intelligibility, the condition for the rational apprehension of the physical world.

In an early point of agreement between Chomsky and Foucault in their 1971 debate, Chomsky noted that Newton's advancement upon the corpuscular philosophy brought with it new rules for the reasonable construction of scientific explanation.[24] Similarly, Descartes' advancement upon scholastic philosophy brought a new way of understanding phenomena, such as thinking and the wilful movement of limbs, which seemed unexplainable by the scholastic principles of action. Chomsky suggests that such phenomena compelled Descartes to postulate a second substance in order to account for that which was unaccountable by the purely material substance:

> The move of Descartes to the postulation of a second substance was a very scientific move; it was not a metaphysical or an anti-scientific move. In fact, in many ways it was very much like Newton's intellectual move when he postulated action at a distance; he was moving into the domain of the occult, if you like. He was moving into the domain of something that went beyond well-established science.[25]

Descartes' concept of the cogito, and the separation of mental and corporeal substance within which it settled like a habitat, was an alien concept that remained incompatible with the scholastic picture of nature; and similarly, Newton's empiricism introduced a foreign and impossible concept of action at a distance into the corpuscular philosophy.

From the perspective of corpuscular philosophy, the reasons for condemning action at a distance are quite clear. For Descartes, of course, matter is continuous and identical with extended space – corporeal substance is the universal plenum of *res extensa*. But Newton's 'active powers' imply *discontinuous* space, since they are active across distances. Consequently, *res extensa* cannot possibly harbour 'active powers', which would be other than strictly material, if not perfectly immaterial. But we realise here the severity of the incompatibility between action at a distance and corpuscular philosophy. There is literally no room in corpuscular philosophy for action at a distance, since there are no distances in matter. Action at a distance is therefore incompatible with the corpuscular concept of matter itself. To make matters worse, Newton's active powers not only offended the picture of *res extensa*, they also disturbed the concept of mental substance. After all, while the material and the mental are strictly separated in Descartes, the two substances still negatively define each other. In other words, physical phenomena are to be wholly explicable in material terms alone, precisely because they are wholly separated from the mental – the separation between the body and the mind strictly circumscribes physical phenomena within the bounds of *passive* material action by contact. It is in this sense that, within the corpuscular system, material continuity forms the condition of physical intelligibility, so the entire regime of intelligibility of corpuscular philosophy is compromised by the concept of action at a distance. Active powers do not merely introduce discontinuous spatial distances into the concept of matter, for they are *active* powers, and so they introduce a different kind of element altogether into the corpuscular definition of matter as entirely passive. In this fashion, Newton admits *heterogeneous* non-material (following the corpuscular definition of matter) elements into the concept of matter. While Newton's various aethers could be rare and subtle matters, active powers are patently not material (again, in the corpuscular definition) while acting at a distance. This constituted a problem not only for Newton's critics but also for Newton himself. As one commentator asks, do 'the aether's actions have material causes, or are they the effects of a non-material active source? Are the ultimate sources of alchemical and mechanical activity material or non-material?'[26] From this perspective, the accusation of occultism seems almost mild, if not completely appropriate.

Now, since the concept of action at a distance introduces heterogeneous elements into the landscape of natural phenomena, we may recognise within it the concept of the milieu. And further, the heterogeneous elements conjoined in the concept of milieu disrupt the categories of mind and matter, just as the concept of *techne* presented a limit to *episteme* and blurred the division between exact and inexact sciences, disrupting the categories of the natural and the social. In both cases, the source of this limit is the becoming of eventualities – *techne* was a mode of determining uncertain events, and the milieu is the conjunction of heterogeneous eventualities that compose phenomena in space. By admitting heterogeneous elements into his account of natural phenomena, Newton's milieu of distant action departs from the corpuscular categories of knowledge – it brings together the disparate elements of rational material and irrational immaterial active powers and blurs the distinction between the corporeal and the incorporeal, which is why Newton, ultimately, is compelled to admit that 'We cannot say that all nature is not alive.'[27] It is not simply the category of matter that loses its integrity here; the categories of mind, and even of life, lose their strict boundaries: after Newton, we no longer have a concept of matter, but only a concept of milieu.

The Milieu in French Biology After Newton: From Lamarck to Comte

'The French mechanists of the eighteenth century', Canguilhem tells us, 'called "milieu" what Newton had referred to as "fluid." In Newton's physics, the type – if not the sole archetype – of fluid is aether ... insofar as the fluid penetrates all these bodies, they are situated in the middle of it [*au milieu de lui*].'[28] Newton's luminiferous aether is an intermediary between distinct physical bodies such as a light source, for example, and the ocular apparatus that reacts to it.[29] The light, then, is transmitted through the milieu, and the milieu is a vehicle of the forces that act through it. But this is not to suggest a return to the universal plenum of corpuscular philosophy – action at a distance is not 'bridged' by the aether (such continuity would cancel the problem of the milieu) – again, the aether 'transmits' or 'communicates' active powers across distances between bodies. The most important point to focus on here is that the milieu is an intermediary *between two centres* (e.g. light source and eye), which makes the milieu an essentially relative notion.[30] This can perhaps be more clearly seen with regard to Newton's gravitational aether, since gravitational force is a consequence of one body put into relation with another body – in other words, it is a *relational consequence*.[31] The milieu can only be conceived as an environment surrounding a central body, like a bowl of water surrounds a goldfish, by dismissing its essentially relative and intermediary status. Such a dismissal imparts a substantial quality to the milieu, as if it were a substance in and of itself. And it is precisely the conflict between the milieu as an active relative notion and the mechanical (substantial) notion of milieu as a field surrounding a body that is of chief interest in its conceptual crossover into biology through Lamarck and Comte.

When Lamarck spoke of 'milieus', Canguilhem observes, he always spoke of them 'in the plural – by which he expressly means fluids like water, air, and light', and when he designates the complex of actions that operate in the space external to the organism (the milieu in a Hippocratic or environmental sense), he uses the phrase, 'influencing circumstances'.[32] So Lamarck adopts the *mechanical* sense of milieu as a fluid, in the sense that organisms function *mechanically* within certain spatial milieus, like a bird's wings carry it through the 'milieu' of air or like a fish's tail propels it through the 'milieu' of water. For the *complex* sense of milieu, as the relational space in between the organism and its complex environment, Lamarck reserves the term 'influencing circumstances'. Comte, on the other hand, uses 'milieu' to designate not only the fluid within which a body operates, but the 'total ensemble of exterior circumstances necessary for the existence of each organism'.[33] However, while this use of milieu might seem complex, Comte understands 'exterior circumstances' mechanically, as a cluster of mechanical mediums or fluids. Thus, instead of treating the milieu as a relative intermediary *discontinuously* connecting heterogeneous elements between organism and environment, in the *Course of Positive Philosophy* Comte conceives of the milieu as a *continuous function* that took, for its variables (or fluids), 'weight, air and water pressure, movement, heat, electricity, and chemical species – all factors that can be studied experimentally and quantified by measurements'.[34] From Canguilhem's biological perspective, the *Course* seems to consider the qualitative categories of biology as empirically quantitative variables of mechanical functions. Canguilhem contends that while there is a subordination of the mechanical to the vital in Comte's later works, in the early *Course* the physical variables of the environment seem to completely subsume the organism, to the point that the biological organism is reducible to the physical functions that support it.[35] Thus, instead of considering the milieu as a site of heterogeneous composition, the early *Course* considers the milieu as a homogeneous function of quantifiable variables.

This brings us to an important dimension of the biological use of the term that is acquired from Comte's early usage of 'milieu'. Whereas Lamarck's 'influencing circumstances' (a term eclipsed by the success of 'milieu') suggests a privileged centre around which influences swarm (or circum-stand), Comte's continuous milieu dissolves the entity into an abstract, homogeneous *extensa* that cannot even be said to surround it:

> The representation of an indefinitely extendible line or plane, at once continuous and homogeneous, and with neither definite shape nor privileged position, prevailed over the representation of a sphere or circle, which are qualitatively defined forms and, dare we say, attached to a fixed center of reference.[36]

In Comte's usage, the milieu is an abstract interface that allows for the reduction of all phenomena into quantifiable variables: 'the now refers to the before; the here refers to its beyond, and thus always and ceaselessly. The milieu is truly a pure system of relations without supports.'[37] It is this abstract prestige of the term that is Comte's contribution, which encouraged French neo-Lamarckians to use the concept of milieu to argue that 'fish do not lead their lives on their own; it is the river that makes them lead it; they are persons without personality'.[38] Thus, without the supports of qualitative differences, the milieu becomes a conceptual instrument for a one-way determinism of the organism through physical variables, fashioning the organism as uncreative, mechanical automata. We have thus returned to a picture of *res extensa* and Cartesian animal-machines. The concept of milieu once again finds itself between two poles: by identifying itself with material variables, it gains all the quantifiability of those variables; in turn, however, the relative and qualitative differences (or the active heterogeneities) of the milieu are dissolved precisely to the extent that it is identified with material variables, rendering it merely a piece cut out from the homogeneous universal plenum. Consequently, Canguilhem finds that from 'Galileo and Descartes on, one had to choose between two theories of milieu, that is, between two theories of space: a centered, qualified space, where the mi-*lieu* is a center; or a decentered, homogeneous space, where the *mi*-lieu is an intermediary field'.[39]

The concept of the milieu, then, is faced with two possibilities. It can either maintain the homogeneity of the natural, thereby separating the natural from the artificial and maintaining the traditional categories of knowledge, which separate the material from the mental and the rational from the irrational; or it can dismantle those traditional categories and admit of irreducibly qualified, relative, active and heterogeneous elements. Lamarck effectively chose the latter through his notion of 'influencing circumstances' since the milieu did not act directly upon the organism, but at a distance, as it were: 'it is via the intermediary of need, a subjective notion implying reference to a positive pole of vital values, that the milieu dominates and compels the evolution of living beings. Changes in circumstances lead to changes in needs; changes in needs lead to changes in actions.'[40] To be sure, this is not to suggest that the milieu does not have a physical dimension, but rather that the constitution of the biological milieu cannot be the object of a materialist reduction. Thus, Lamarck's 'influencing circumstances' continues the trajectory that Newton advanced by installing active powers within the milieu, a manoeuvre that also prohibited the wholesale subsumption of the natural into the material. And it is this legacy of the milieu that Canguilhem endorses:

> Of course, we might still speak of interaction between the living and the milieu even from a materialist point of view – between one physic-chemical system cut out from a larger whole, and its environment. But to speak of interaction does not suffice to annul the difference between a relation of the physical type and a relation of the biological type.[41]

To be sure, the reduction of nature to the lowest common denominator holds little promise for the study of life with its irreducibly qualitative phenomena – the worlds of organisms are populated by affects, not colliding particles, and organisms do not merely react to stimuli, they evaluate and respond – thus we cannot compose the organic order out of elementary physical mechanisms. In short, biology requires an approach of relative, qualitative, asymmetrical construction, where organism and milieu operate alongside each other in order to conduct a complex sequence of events.[42] But more profoundly, by insisting on the heterogeneous, relative and active dimensions of the milieu, we find that the concept of milieu resists a hierarchical structure of nature *tout court* (such as Comte's, which escalated from physics to chemistry and biology, culminating with sociology), and encourages a real heterogeneous and relative composition of eventualities, beyond any disciplinary structure of knowledge that isolates distinct categories such as the material and the mental.

Artificial Life, or, the Role of Milieu and Technicity in the Eventual Composition of Organic Life: von Uexküll and Leroi-Gourhan

Jacob von Uexküll, in his *Foray into the Worlds of Animals and Humans*, presents a strong case for the qualitative and relative composition of organism and environment. His description of a tick's *umwelt*, or worldly environment, remains one of the most memorable illustrations of the life of an insect.[43] The tick, after all, leads a very simple life. At one to two millimetres long, it climbs up some shrubbery and waits for its mammalian prey, which it discerns by smelling the butyric acid in its sweat. Once it detects its prey, it either falls on top of the mammal or lets itself be brushed off by it. Then it scurries to the least hairy spot, which it finds through its sense of touch, and bores its head completely into the mammal's skin and begins to pump a warm flow of blood into itself, swelling to the size of a pea.

What is interesting about this is that out of the theoretically infinite sensory possibilities that the world has to offer, the blind and deaf tick initially responds only to the smell of sweat – 'just as a gourmet picks only the raisins out of the cake, the tick only distinguishes butyric acid from among the things in its surroundings'.[44] And further, if a drop of butyric acid is placed on a metal table underneath the tick, it will drop but, sensing cold metal instead of warm skin, it will climb back up to its outpost and resume its patient waiting. Neither does the tick have any judgement of taste, for it will pump any warm liquid into its body once it bores its head past a skin-like membrane.

Beyond the Nature-Artifice Divide

Here is a composition of a world in three strokes: a series of *sense-events* of touch, smell and temperature. Of course, the tick does more than sense, it also effects, but the composition of its life in the form of *effect-events* simply follows the sequence of sense-events – if it smells butyric acid, *then* it drops; if it touches the mammal, *then* it searches for skin; if it touches skin, *then* it bores; if it senses warm blood, *then* it slowly begins to pump. If any sense-event in the sequence does not produce the effect-event, then the sequence begins anew. And the tick can lie waiting for a sense-event for an incredible eighteen years without any effect-event being triggered.[45]

'Figuratively speaking', von Uexküll says, 'every animal subject attacks its objects in a pincer movement – with one perceptive and one effective arm.'[46] That is to say, each animal subject *perceives* only that which interests its organic constitution, and it moves through the sequence of its life with *effective* techniques by which it attempts to conduct the events that interest it. The worldly milieu is thus relative to the animal, composed with regard to its qualitative selections of percepts and its technical actions. In this sense, the space between humans and ticks is relative to their perceptive and effective capacities. The human organism does not exactly share a common world with ticks; rather, we may say that the human and tick *intervene* upon each other's relative eventualities. And the relativity of these eventualities entails that there is no overarching *umwelt* that contains all the animal *umwelten*.[47] The problem of the body and the environment becomes a temporal and technical problem (intervening upon relative eventualities) rather than a substantial and spatial one (sharing a world); it is a question of techniques for perceiving and effecting.

Leroi-Gourhan's analyses of anatomical technicity in *Gesture and Speech* allow us to conclude that humans, while of course different from ticks, are not exempt from this relative and eventual composition of organism and environment. Arguing against the myth of the ancestral monkey, Leroi-Gourhan discusses the discovery (in Kenya in 1959) of the remains of the *Zinjanthrope* (*zinj* for East Africa), an adult *Australopithecus* (Latin *australis* for 'southern'; Greek *pithekos* for 'ape') accompanied by stone implements. The *Australopithecus* enjoyed an upright gait, a face with a relatively retreated mandible, and a very small brain. The myth of the ancestral monkey, on the other hand, conceived of walking upright as a largely cerebral initiative – the brain enlarging and becoming more and more complicated until, one fine evolutionary day, a monkey 'thought' to walk upright and use tools. The discovery of *Australopithecae* with small brain-cases provided compelling evidence that the brain was not the cause of erect posture but rather its beneficiary.[48] It was a development of our anatomical technicity, Leroi-Gourhan argues, which conditioned the subsequent cerebral development that required the protrusion of the skull through a forehead: mechanically, an upright posture encourages the recession of the mandible, which in turn encourages the protrusion of the forehead (a protruded mandible entails a longer face with a brain pan barred by the orbital ridge, as we see in the monkey with its hunched over posture).

But the important factors in this evolutionary story are the coincidental freedoms that the head, mouth and hand acquire through bipedalism. The hand, for example, is freed from its locomotory assignment by an upright gait, and it can thus be used primarily for digital manipulation, while the mandible (and the mouth in general) is freed from its specialised task of reaching for and grasping non-manipulated food. Leroi-Gourhan admires Gregory of Nyssa's *Treatise on the Creation of Man* in this regard:

> If man had been deprived of hands, his facial parts, like those of the quadrupeds, would have been fashioned to enable him to feed himself: his face would have been elongated in shape, narrow in the region of the nostrils, with lips protuberant, horny, hard, and thick for the purpose of plucking grass.[49]

Of course, the ramifications of bipedalism reach far beyond the composition of the human face. The freeing of the hand from locomotion allows for the emergence of complex manual and digital operations required for tool-making, and the mandible retreating allows for the articulation of the tongue as the forehead protrudes. Gregory of Nyssa continues: 'If our body had no hands, how could the articulated voice form inside it? The parts around the mouth would not be so constituted as to meet the requirements of speech. In such a case man would have had to bleat, bark, neigh, low like the oxen, or cry like the ass, or roar as the wild animals do.'[50] In other words, we cannot understand the hand, the mouth, the face and the head in isolation – they are consequences of *coincidental freedoms* that emerge in relation to one another.

The freeing of the hand from its locomotory assignment so that it can gesture and digitally manipulate is an instance of what Leroi-Gourhan calls *de-specialisation* or *generalisation*. The mouth-mandible structure is similarly freed from its specialised function for plucking food and gains the capacity for articulation, and with it the head is freed for the concentration of sensorial orifices in the face (these freedoms from specialised assignments are also instances of what Deleuze would later refer to as *deterritorialisation*).[51] All a result of bipedalism, these de-specialisations culminate in the emergence of an *anterior field* composed through the heterogeneous coordination of the facial and the manual poles that resulted from bipedal development.[52] This anterior perceptual field draws not only a 'front' and a 'behind' of the body, it constitutes, rather, an entire directional space constituted by perceptual and effective events. The eye and the hand share a destiny in the human, in the sense that what the eye sees is related to what the hand manipulates, and so on – the sensory features of humans do not exist in isolation, but compose the anterior field through their heterogeneous coordination. In this fashion, the anatomical technicity of bipedalism ultimately determines the *effective* space and time of human beings through the composition of the anterior field. The elementary construction of human environments is thus a relative consequence of such perceptive and effective technologies. Far from being exclusive to the tick, the eventually constructive and relative dimension of the milieu is equally operative in the human.

Of course, the development of various technologies, from tool-making to orality and literacy, elaborates the human milieu considerably, 'initiating a long transitional period', Leroi-Gourhan writes, 'during which sociology slowly took over from zoology'.[53] The evolved hand becomes available not only for complicated digital operations, but for gestures to accompany the vocal utterances of the mouth, which in turn has become available for speech alongside the development of the brain. While the *Australopithecae* seemed to have possessed their stones in fingers like animals have claws, as if they were mere anatomical extensions, we enjoy a certain distance from technologies, in the sense that our tools are comfortably detached from our body, not only spatially but temporally (since we use tools only when the occasion arises). Leroi-Gourhan attributes this to the fact that, as previously mentioned, our evolution was one of generalisation, evolution in the opposite direction from specialised animals: 'we run less quickly than the horse, cannot digest cellulose like the cow, climb less well than the squirrel . . . our whole osteomuscular mechanism is super-specialised only in remaining capable of doing all of those things'.[54] Other animals persisted upon an *internal* and specialised track of evolution, while human evolution is characterised by an *externalisation* of technology – instead of a stomach evolved for the task of eating specific foods, we make fires and ovens, instead of a clawed feet and hands, we make shoes, ropes and spikes in order to climb trees, etc. So while our advanced technologies significantly extend our abilities, our most basic tools are a testament to how poorly adapted our mere body is for the challenges of its (natural) environment – clothing, shoes and cutlery are, in a sense, indications that our bodies have become de-specialised.

Furthermore, it is paradoxically our generalised or de-specialised evolution that conditions the richness of human consciousness – the distance we have from our techniques allows us to select from among them, a process of selection characterised by conscious deliberation toward a goal. This is perhaps the most profound question for Leroi-Gourhan: how conscious deliberation of events is coordinated by techniques. The most fascinating aspect of Leroi-Gourhan's evolutionary story is that the distance between the incorporated technologies of *Australopithecae* (their stones as mere extensions of their anatomical technicity) and the detached technologies of *Homo sapiens* (their clothing as prosthetic skin, their fires and stoves as prosthetic stomachs, their axes as prosthetic beaks or claws) is a distance produced by a series of steps *installed within technological sequences themselves*: we cannot practically separate consciousness from technics. Discussing this point, Leroi-Gourhan shows how the historical stages of technology betray memorial and imaginative advances – from the bi-faced stone whose two-strike construction requires only a two-step memory to the flakes and knives of later periods that require a sequence of steps that must be memorised for their construction. Leroi-Gourhan's basic point is that the eventual sequences of our instruments parallel the eventual capacity of our memory: our memory becomes more and more elaborate *alongside* our instruments. The bi-faced stone requires a mere two-step construction, implying a two-step memory, while the knife and the axe require significantly more steps and consequently imply a more elaborate memory (in addition to a more elaborate sense of the future and the anticipated use of the instrument). Furthermore, in the case of tools like the axe there must be an *imagined* result that no longer even resembles the original working material of stick and stone, indicating a level of abstraction within technical syntax (the bi-faced stone, on the other hand, looks like a stone throughout its construction). In this fashion, Leroi-Gourhan discerns a *parallel*

Beyond the Nature-Artifice Divide

evolution between so-called intellectual capacities and technological advancement, between milieu and technics.[55] The crucial point here is that there is no way to separate consciousness from environment – they are mutually composed such that consciousness itself may be considered as a technical activity: from the bi-faced stone to knives and axes, and ultimately all the way to orality and literacy, we think through our technologies. Small wonder, then, that our concepts of nature have always been formulated in terms of our technologies – reason is no more independent of the sequences of corporeal life than the knife is independent from the anatomical movements and milieu within which it has its use and intelligibility. The distinctive feature of human evolution is the de-specialisation or externalisation of technology with the simultaneous evolution of reason built into those very technical operations.

Regarding massive technological advancements such as orality and literacy, Leroi-Gourhan finds that they do not merely indicate an extension of human abilities, but rather an entire re-ordering and reconceptualisation of humans and society.[56] Similarly, alongside the development from techniques of hunting to the techniques of stockbreeding there emerge the artifices of new social orders.[57] The significance of the correlation between technological development and social development is thus immense – from migrating hunting societies to sedentary stockbreeding and agricultural societies, there is an entirely new arrangement of bodies, with entirely new hierarchies, codes of alliances and filiations and modes of deliberation. In a largely overlooked early work, Deleuze presents a thesis that seems indebted to Leroi-Gourhan, claiming that institutional artifices 'impose on the body, including its involuntary structures, a series of models, and provide intelligence with a kind of knowledge – a possibility of foresight in the sense of a project'.[58]

From these considerations, we may be led to conclude that anatomical technicity is extended and transformed by so-called 'artificial' technologies that continually rearrange the distinctive eventualities of human beings. Yet, from those same considerations, we can no longer presume a 'natural' human being who eventually thought 'artificially' and began to make tools. The organism is *already* artificial – equipped with an anatomical technicity characterised by ways of perceiving and ways of effecting marks upon the milieu, both of which are artificial and compositional, albeit unconscious. The distinction here is thus not between natural and artificial technicity, but between conscious and unconscious technicity. The human being, just like the tick or the crab, is invested in its milieu artificially, which is simply to say that techniques conduct events (and not the other way around). What is distinctive about the human milieu is the detachment of technicity from its *specialised* assignment, as is the case with the *generalisation* of the hand, mouth and head, because, for Leroi-Gourhan, this detachment allows for the hand, mouth and head to be *re-assigned* for gestures, speech, deliberation, etc. And it is this re-assignment that, in turn, constitutes a field of practical rationality within the composition of sequences of events. Consequently, *practical* rationality is precisely the rationality that is produced and facilitated through the artifices that conduct events. In this fashion, the composition of the human milieu is technical – human eventuality unfolds within the relative composition of milieu and technics. The essential point here is that from the perspective of eventuality, the natural human being *is thoroughly artificial*, or perhaps better, that the basic zoological constituents of the human, whether anatomical or physiological, cannot be understood in terms of a distinction between the natural and the artificial. In fact, the natural itself is already artificial, not only in the sense that the concept of nature is defined technically, but also in the sense that the events that constitute the distinctions between the organic and the inorganic are no less technically conducted than the distinction between the conscious and the unconscious, in the very basic sense that both are determined by the composition of a series of events. From the perspective of time and events, the set of disciplines and objects understood as human or social cannot be defined independently from the set of disciplines and objects understood as natural. By refraining from distinguishing the natural from the artificial, we obtain a more integral perspective on both, acknowledging that the human is not only an eventuality relative to nature, but that artifice itself introduces or expels eventualities from milieus. That is to say, once we consider the complex of events that produces the distinction between the human and nature, there is no way to identify that complex as either human or natural. We have already seen that there is no overarching *umwelt* but only a relative composition of events called milieus. There remains, however, a composition of eventualities at work behind the production of human-nature milieus, arranging the technologies and conducting the events. The eventualities, however, cannot be understood simply in terms of events – it is not a question of the artifice that produces the nature-human dichotomies. Eventualities must be understood in their own dimension as processes that produce events, as processes that produce artifice and technology, but as processes that are nonetheless integral to the events and technologies that they produce. There are, as it were, machines that both compose and are composed by technologies.

Machine

MACHINES AND EVENTUALITIES

While Foucault's concept of technology, as practical rationality, hinged upon consciousness, Leroi-Gourhan's concepts of technics and milieu covered both conscious techniques, such as the production of an axe or a stockyard, as well as unconscious technicity, such as the production of an anterior field through the heterogeneous coordination between the eye and the hand. But just as the milieu of the tick is different from the milieu of the hummingbird, the snail and the human, so there are different times or arrangements of becoming that are produced through the respective eventualities of such creatures. When discussing the time of *conscious* technology, I have employed the term 'event', reserving the term 'eventuality' to denote the temporality of *unconscious* technicity. But this raises another question, namely, if we grant that technology conducts events, and not the other way around, then how do technologies and events arise in the first place?

This question asks for the genetic conditions of technology and events, and to some extent the concept of milieu serves as an answer – the milieu is the space and time composed through the heterogeneous eventualities, that is, a space and time produced through the conjunction of various elements across various disciplinary registers (physical, chemical, biological, social, etc.). But how are these heterogeneous eventualities produced? And what are the conditions that facilitate their conjunction? Deleuze and Guattari's concept of machine furnishes us with a promising approach to these questions precisely by addressing the unconscious register of what I have been calling eventualities, a term which thus far has been used without explication. This section outlines how the concept of machine reconfigures the concept of milieu, and how the machinic theory of milieu involves two temporal axes, a conscious time of technical conduct and an unconscious time of machinic production.

The Concept of Machine in General

The cover-leaf of *Anti-Oedipus* bears a painting by Richard Lindner, *Boy with Machine*, depicting, as the title suggests, a boy embedded within a complex of gears, pulleys, levers and cartridges. It illustrates the situation of little Joey, a child 'who can live, eat, defecate, and sleep only if he is plugged into machines provided with motors, wires, lights, carburetors, propellers, and steering wheels: an electrical feeding machine, a car-machine that enables him to breathe...'[59] Little Joey functions, we might say, within a machinic assemblage that constitutes the milieu of his desire. In a way, these are technical machines, in the aforementioned sense of *techne*, since they are part of a practical rationality oriented toward the conscious goal of eating, for example. Of course, this is no ordinary technical apparatus. In fact, the machines would sometimes 'explode' and he would scream 'Crash, Crash! . . . hurling items from his ever present apparatus – radio tubes, light bulbs', until one of the thrown objects would shatter, at which point 'he would cease his screaming and wild jumping and retire to mute, motionless nonexistence'.[60] So it is difficult to distinguish where little Joey 'ends' (as an operator) from where the assemblage of technical machines 'begins' (as operated), at least without tearing little Joey apart, as it were, since the machines are required for his functioning: he is quite literally plugged into his concrete environment of actions and events through a machinic assemblage – his desires are machinic, he is himself a complex of machines.[61]

As Daniel Smith notes, Deleuze and Guattari's concept of desiring-machines is something of a resolution of Freud and Marx on the question of the unconscious origin of our conscious desires.[62] Basically, for Freud, the foundation of desire was *libidinal economy*, some internal drama hidden in the deep recesses of the psyche; whereas for Marx, the source of desire was *political economy*, emerging from class-consciousness and the socio-political relations external to the subject. If desire begins with the individual unconscious, it must nonetheless go through the turbulence of social integration; on the other hand, if desire begins with social structures, then those social structures must nonetheless reach quite far, extending to infantile development. What Deleuze and Guattari accomplish, through their concept of desiring-machines, is to draw a line perpendicular to the opposing starting-points of Freudianism and Marxism: libidinal economy is *already* political economy: 'social production

Beyond the Nature-Artifice Divide

and desiring-production are one and the same'.⁶³ Desire is located neither 'inside' nor 'outside' a subject, but rather is *produced* by an assemblage of machines, which traverse the entire milieu of the subject, as little Joey is embedded within a complex of machines that render him functionally continuous with his environment.

The difficulty of distinguishing between operator and operated in the case of little Joey brings us right to the heart of the problem between technology and machines. Foucault repeatedly emphasised the correlation between technology, knowledge and society. We can take, for example, lepers in the Middle Ages – where we find a technology of *exclusion* that produces knowledge of lepers – the question was 'who are the lepers?' The answer to this question was produced through a technology of separating them through laws and sets of religious rituals, which produced the knowledge of lepers: the lepers are those that are banished, excluded from the community. This technology of exclusion becomes one of *quarantine* with the sixteenth-and seventeenth-century problem of the plague: in order to identify who were the plagued, we find a technology of inspection, regulations indicating where people can go, requiring them to present themselves to inspectors, an entire disciplinary system oriented toward finding out where the plague was, who were the plagued, and subsequently detaining them within the town. Things change with the outbreak of smallpox in the eighteenth century, when the question becomes 'how many people are infected with smallpox, at what age, with what effects, with what mortality rate, lesions or after-effects, the risks of inoculation, the probability of an individual dying or being infected'; in other words, the problem is no longer one 'of exclusion, as with leprosy, or of quarantine, as with the plague, but of *epidemics* and the medical campaigns that try to halt epidemic or endemic phenomena'.⁶⁴ These examples highlight the historical correlation between technology (exclusion, quarantine, epidemics and medical evaluation) and knowledge (Who are the lepers? Where is the plague? What is the mortality rate, etc., of smallpox?), which informs the social order (how bodies and individuals are identified and arranged within a social space). The basic point here is that *regimes of knowledge are produced through forms of technology, which, in turn, are intelligible only from within the context of the social orders that are informed by them.* Similarly, anthropologists like Leroi-Gourhan often describe the connection between techniques and an ensemble of social and environmental factors: we cannot understand a stockyard outside of a stockbreeding society, with all that entails (a sedentary order, a certain distribution of tasks, principles of filiation and alliance that support that set of tasks, etc.). Marx, of course, had already made this essential point in *Capital*: 'Technology reveals the active relation of man to nature, the direct process of the production of his life, and thereby it also lays bare the process of the production of the social relations of his life, and of the mental conceptions that flow from those relations.'[65] There is little need, then, to say we all have technologies through which we relate to nature, and that these technologies garner their intelligibility from the social formations within which they operate.[66] Of course, Deleuze and Guattari's concept of machine does not simply reiterate the plain fact that there is a connection between nature, technology and society; rather, it identifies the *same process* at work in producing both: 'man and nature are not like two opposite terms confronting each other . . . rather, they are one and the same essential reality'.[67] In other words, instead of starting with the social and the natural as the two poles of an elementary confrontation, Deleuze and Guattari focus on the process that produces the very distinction between the social and the natural, revealing the artificiality of the distinction itself.

Machines, Flows and the Synthetic Process of Production

The identification of man and nature within machinic process is facilitated by the essentially connective nature of machines themselves; that is, machines are non-metaphorically synthetic – connecting and disconnecting – as an infant's suckling mouth-machine connects to the mother's breast-machine to produce the eventual flow of milk, or as the mouth-machine can also produce sound through the larynx-machine tapping the flow of air, and so on. But this is not to suggest that machines are substantial entities; in fact, they resist reification by their essentially connective nature.[68] Little Joey is certainly embedded within a machinic assemblage, but the *machinic* dimension of the assemblage is not constituted by the stand-alone objects within which he operates. In order to examine this point, we may briefly explore the relation between the concept of machine and its correlate concepts of flow and synthesis.

Following biologists Umberto Maturana and Francisco Varela, Guattari defines a machine 'by the ensemble of interrelations and its components, independently of the components themselves'.[69] In other words, the relational syntheses that obtain among the components

Beyond the Nature-Artifice Divide

constitutes the machine, not the componentry itself. For example, the mouth-machine is constituted by its synthetic relation to other machines, and so it is only a machine in so far as it is engaged in the machinic production of a flow, such as the flow of milk or the flow of sound. The term 'flow' here can be understood in an economic sense, which in turn illuminates the productive aspects of machines. Take, for example, the flow of money as it is distributed throughout a standard arrangement of industrial capitalism: the owner of a factory or some means of production employs a labourer to extract labour and produce surplus, augmenting the owner's capital while merely preserving the labourer. The labourer, finding the situation unfair, may purposefully go on strike, disconnecting from the flow of money, but the capitalist may be in a position to simply wait, subsisting on her capital. After all, the labourer has less capital, and must eventually return to work in order to reconnect to the flow of money, to survive and perhaps to provide for dependants. So far this is a familiar story, but the *machinic* point is that this arrangement works only because the labourer and the capitalist *both believe in money*. In other words, the flow of labour, the flow of money and the flow of ownership requires an investment in the full arrangement of capital. And for Deleuze and Guattari, this investment in capital does not arise from an internal psyche or the external social structures outside of the individual, and neither is it a merely epistemic belief in money. There is nothing about money or capital in itself that makes it valuable, rather, the entire system or assemblage of capital carves out an economic reality within which certain subjects like owners and labourers can be identified and play out their relative destinies. It is in this sense that investment is constituted by an arrangement of desiring-machines, in the sense that flows of labour, money and ownership are produced and facilitated through machinic assemblages. Similarly, the desire to be a lawyer or an academic entails the entire machinic assemblage within which those subjects have their currency (law courts, universities and all the other institutions that make that desire possible). The essential point here is that conscious desire for money is a *state* of machines produced through their synthetic assemblage. Simondon makes a similar point when discussing the assembly line in *On the Mode of Existence of Technical Objects*. We normally think that the assembly line introduced standardisation into the workflow of factory labourers. But Simondon argues that it is more adequate to say that standardisation was a *principle* of technology that drove the production of the assembly line – the assembly line is not the cause of standardisation but rather one of its effects. In fact, we do not standardise production through the assembly line so much as the formation of stable and standard types facilitates industrialisation: 'It is not the production-line which produces standardization; rather it is intrinsic standardization which makes the production line possible.'[70] Further, this intrinsic standardisation entails that we are standardised ourselves, in the sense that a whole range of technologies are deployed to shape our days as well as our sense of work and life – technologies that have us wake up and go to work at a certain hour, at a certain pace, with lunch breaks at noon, etc. Society does not produce technologies to accomplish its own internal aims, rather, there is something intrinsic to technology that guides its own evolution. Thus, instead of considering technologies as mere utilities that help us perform tasks, Simondon prefers, then, to interrogate technological principles that transform the composition, not only of technology, but also of our social lives. This is a question that moves toward the concept of machine, which traverses the categories of nature, subject and society. Now, Deleuze and Guattari are very careful not to simply reiterate the fact that a wider social order is implied when considering a subject or desire – crucially, machines are neither individuals, nor institutions, nor societies (the concept of machine endeavours to explain the production of these categories, it does not assume them). Since machines are not defined by scale or category but rather by flows or their synthetic function, it makes no difference if we consider vast social machines such as the internet or small desiring-machines such as the eye that consumes its images. This trans-disciplinary and cross-categorising capacity of the machine is due, in part, to the fact that the concept of machine is not a spatial concept but rather a temporal one.

The term 'flow' highlights the temporal orientation of machines, since flows of money, flows of milk and flows of sound are not so much substantial entities as moments or products of a machinic process. The basic temporality of the concept of machine is also the reason why machinic process cannot be understood as a law of nature or some abstract elementary process, since the machines themselves are constituted by syntheses and flows. A tick is a machine when intervening upon the flow of blood, but the flow of ticks cannot be abstracted from the flow of mammals alongside which their eventualities unfold. Just as technologies are unintelligible apart from the social orders that facilitate them, machines are indistinguishable from their connections with other machines and flows within an assemblage. Thus the concept of machine is thoroughly concrete, temporal and trans-disciplinary – as it is composed not by substances but by connections which in turn condition the identification of substances, the concept of machine dissolves the distinction between subject and object and disrupts the traditional categories of knowledge that govern the philosophy of nature.

Production and Connective Assemblages

Deleuze tells us that an assemblage 'is a multiplicity which is made up of many heterogeneous terms and which establishes liaisons, relations between them, across ages, sexes, and reigns – different natures. Thus, the assemblage's only unity is that of a co-functioning: it is symbiosis, a "sympathy".'[71] In Stoic philosophy, 'sympathy' referred to the interconnectedness of parts and wholes such that when one part changes the whole is likewise affected. The use of 'sympathy' to designate the connection between diverse elements across the cosmos is maintained throughout Renaissance, sixteenth- and seventeenth-century natural philosophy, and the term was sometimes used when treating topics such as optics and magnetism, which often relied on 'occult' qualities for their explanation. But Deleuze's reference to 'sympathy' here does not merely suggest a return to occult qualities, rather, it nominates the assemblage as a machinic theory of milieu. Thus, Deleuze and Guattari reformulate the problem of the milieu (how heterogeneous elements such as cities, dews, humans, stars, carburettors and wars, for example, are all linked together) into a problem of machinic assemblages (how are machines attached and detached, or *synthesised*, so as to produce or interrupt a flow of dew, of starlight, of sound, war, etc.). This machinic reformulation of the concept of milieu recasts the heterogeneous elements of the milieu within a synthetic process of production beyond the disciplinary hierarchies of knowledge (the physical, chemical, biological and sociological, or the natural and the social, the mental and the material, etc.). To be sure, such a reformulation demands some experimental terminology that does not enjoy the familiar clarity of traditional categories. Beyond these categories of knowledge, Deleuze and Guattari describe milieus as 'vibratory blocks of space-time' that communicate asymmetrically 'from the inorganic to the organic, from plant to animal, from animal to humankind, yet without this series constituting a progression'.[72] While the characterisation of a milieu as a 'vibratory block of space-time' might seem unruly, the upshot of the experimental terminology – namely, the absence of a progression from plant to humankind – is a promising endeavour for a reformulation of the concept of milieu. In order to disrupt the traditional disciplinary hierarchy of knowledge, the progression from lower-level to higher-level functions needs to be abandoned; not only because we hazard falling back into an unproductive and reductive picture of the natural sciences (where biology is reduced to chemistry, chemistry to physics, etc.), but also because we hazard falling back into a picture of basic processes of nature (the subject of natural sciences) over and against human practices (the subject of social sciences), as if nature was lower level material for the higher level operation of human activity. As we have mentioned repeatedly, we have always understood nature alongside technologies, and the history of the concept of nature cannot be understood apart from the history of technology and the distinctive evolution of the human milieu through technics. For these reasons, Deleuze and Guattari do not postulate two registers of reality, a higher level register of technologies, cities, dews, humans and wars, and a lower level register of differential machines and flows that produce them. There is only a difference of *regime* within the same process of production. But, in order to examine this point, we must distinguish two temporalities at work in technologies and machines.

Events and Eventualities

We might approach the distinction between these two temporalities by examining two different ways of narrating history. First, we have what we might call a *horizontal* history that overlays becoming with a catalogue of dates, names and places, arranging them in a successive order, as historians do, through a procedure of providing evidence, of arguing why what happened occurred when it did. We might call this horizontal history *official* history, or history in the form of a time-line. This kind of history, however, only comprehends the event as it is actualised, that is to say, it equates events with states of affairs. Foucault admires Deleuze's evaluation of the event because it avoids the trappings of this kind of official history of states of affairs, because for Deleuze, the event does not designate a substantial state of things, rather, the event lies on an entirely different register (although I hasten to add that we use the term 'eventuality' to refer to what Deleuze would call an 'event', reserving the term 'event' to designate the extensive state of affairs that nonetheless productively

Beyond the Nature-Artifice Divide

features in differential process, albeit as a product of that process). Horizontal history, on the other hand, takes the event as a state of affairs, 'something that could serve as a referent for a proposition'.[73] Charles Péguy, in *Notre Jeunesse*, writes that history will always let us know 'the grand drama, the stage, the spectacle', but

> what we want to know is what went on behind, below, beneath the surface, what the people of France were like; in fact, what we want to know is the *tissue* of the people in that heroic age, the texture ... What we want is not a Sunday version of history, but the history of every day of the week, a people in the ordinary texture of its daily life; working and earning, working for its daily bread, *panem quotidianum*; a race in its reality, displayed in all its depth.[74]

Horizontal history provides us with a time line of facts, events and states of affairs, but it covers over the depths beneath them, the sub-representational forces of becoming that produced the state of affairs. Horizontal history is a history narrated by clock time, which understands everything – whether a technical object, an organism or a historical event – mechanistically, as if it were a matter of material configurations moving through instants of time. But it remains difficult for horizontal history to account for the emergence of the new and for the trans-disciplinary connections that constitute it. Consequently, we require another kind of history, what we might call a *vertical* history.

In 'Theatrum Philosophicum', Foucault writes that

> To consider a pure event, it must first be given a metaphysical basis. *But we must be agreed that it cannot be the metaphysics of substances, which can serve as a foundation for accidents; nor can it be a metaphysics of coherence, which situates these accidents in the entangled nexus of causes and effects.* The event – a wound, a victory-defeat, death – is always an effect produced entirely by bodies colliding, mingling, or separating, but this effect is never of a corporeal nature; it is the intangible, inaccessible battle that turns and repeats itself a thousand times around Fabricius, above the wounded Prince Andrew.[75]

Foucault's statement regarding Deleuze and Guattari's theory of the event here brings us to the essential problematic of vertical history. I'd like to approach this problematic somewhat inexactly by reference to Alessandro Portelli's oral histories project, because oral history 'takes place', as it were, between the present and an ever-changing past, moving between the story-teller and the listener. Oral history is rejected by official history because it is thoroughly washed with 'errors' that range from innocently mistaken dates to completely imaginary transpositions and fabrications. But as Portelli shows, these errors come to form precisely the merit of oral histories, because they reveal not only the dreams, desires, prejudices and myths of the narrators, but through their transformative retellings they also reveal the way that events are carried throughout time, the way they unfold like a memory, like a wound. In *The Death of Luigi Trastulli*, for example, Portelli informs us of a historical record that testifies that 'Luigi Trastulli, a 21-year-old steel worker from Terni, an industrial town in Umbria, central Italy, died in a clash with the police on 17 March 1949 as workers walked out of the factory to attend a rally against the signing of the North Atlantic Treaty by the Italian government.'[76] The oral history of this event strays considerably and significantly from this record. In some versions, he is a martyr for the workers protest movement, in others, a young and dedicated family man who was rushing home; he dies either nailed against a wall high above the ground, or shot and run over by a jeep driven by a Sicilian; the date of his death stretches from the NATO demonstrations in 1949 to the steel workers uprising four years later in 1953, and so on. So the stories make use of the horizontal history of dates, but liberally and creatively. The stories thus take place and resonate a within a social memory of the death of Luigi Trastulli embedded within a complex of a community's political and economic triumphs and disappointments. The errors, then, are telling because 'beyond the event as such, the real and significant historical fact which these narratives highlight is the memory itself', Portelli says, 'the causes of this collective error must be sought, rather than in the event itself, in the meaning which it derived from the actors' and narrator's state of mind at that time; from its relation to subsequent historical developments and from the activity of memory and imagination'.[77] What we might say here is that oral history is an art whose material is history itself – it revolves around the question of the *use* of history; in fact, it is a *techne* of history replete with levels of prosody and fabrication, a technology not only for dealing with the past but for intervening in the present, as the retelling of such stories forms a technology of memory oriented toward treating a social wound by avoiding, at all costs, 'a message of collective powerlessness and defeat'.[78]

In Portelli's words, the problem with what we've called horizontal history is that 'all sorts of events happen simultaneously at any given moment, and the building of a chronological paradigm implies a selection of homogeneous events from among those happening at any given time'.⁷⁹ What is at stake here are two different conceptions of history and the event facilitated by two different technologies of managing time. Horizontal history manages time through representation in the figure of a successive and linear discontinuity of instants, that is to say, discrete and homogeneous parts, while vertical history is a non-linear, continuous complex of heterogeneous forces, involving, as it does, heterogeneous communication between utterances, listeners, socio-political regimes, jeeps, guns – in short, re-tellings of utterances and listeners in their concrete socio-political and material ecology.

I have introduced Portelli's oral histories project as an inexact corollary to the problematic of vertical history. Strictly speaking, a jeep driven by a Sicilian, NATO demonstrations and a steel workers' uprising in Umbria are all events (in the horizontal or extensive sense) – they are techniques for narrating history by identifying states of affairs. Nonetheless, the non-discursive problematic that Portelli's project addresses, namely, the problem of becoming, of change, of the production of events and the new, in short, the part of history that resists horizontal stratification or capture by a calendar, is visible in the 'errors' of oral histories. Previously, we considered the term 'event' as relating to *conscious* technology, reserving the term 'eventuality' to denote the temporality of *unconscious* technicity. From the two distinct methods of narrating history, we can now return to these two terms and elaborate further distinctions between them.

In *A Thousand Plateaus*, Deleuze and Guattari locate machinic assemblages within two layers that we will refer to as events and eventualities:

> One side of a machinic assemblage faces the strata, which doubtless make it a kind of organism, or signifying totality, or determination attributable to a subject; it also has a side facing a body without organs, which is continually dismantling the organism, causing asignifying particles or pure intensities to pass or circulate, and attributing to itself subjects that it leaves with nothing more than a name as the trace of an intensity.⁸⁰

The level of 'strata' or 'organism' here is the level of horizontal history, which we have also referred to as clock time – it is the calendar of actualised or determined events. Deleuze and Guattari deploy their notion of the 'body without organs' to discuss the other level of machinic assemblages that 'continually dismantle' the organism, the strata and horizontal history. The term originally comes from Antonin Artaud's 1947 radio play, 'To Have Done with the Judgment of God': 'When you will have made him a body without organs,/ then you will have delivered him from all his automatic reactions/ and restored him to his true freedoms.'[81] In the first chapter of this book we examined judgement in Descartes as a faculty that connects the cogito to the world. Judgement is what determines things, what identifies the barrage of sensations as a 'man' or as 'wax'. What would it mean to have done with judgement – to have done with the faculty that separates the subject and the object and attempts to reconnect them epistemologically, to have done with the faculty that identifies, categorises and classifies the world? Deleuze and Guattari employ the concept of a body without organs to excavate the state of machines (the state we call eventualities) that resists judgement, which resists the codes and disciplinary overlay by which the intelligence apprehends the world. To be sure, the term 'body without organs' has received a lot of commentary, but perhaps the shortest route to the elucidation of the term is to consider an egg. The egg is literally a body without organs – there is yolk and whites, a cytoplasm, but there are, strictly speaking, no organs within the body of the egg. Yet within this body without organs lies the entire *genetic* (I use this word metaphysically) sequence for the production of organs and organelles. One may say that the organic events are latent within the virtual eventualities of the egg like so many axes and thresholds that facilitate the embryonic sequences.

Notwithstanding, one worry that comes with thinking about the body without organs as an egg is the tendency to consider it a self-contained substantial entity and, more, one that may disappear after the development of the embryo. There is no question of the chicken and the egg here. There is, instead, a question of intensities within the extensive, a question of the ecological eventualities within which creatures and trees emerge and take their place. The Calbuco volcano that erupted in Chile in 2015 can be considered a substance only by abstracting from the tectonic pre-history of what became Chile, as well as the epochal thermal and gravitational forces that roiled the igneous and molten lava. The event of the Calbuco volcano erupting was an eventual series within a much longer and wider history of events that conditioned its eruption. But more profoundly, the eventual arc of the volcano's eruption was itself produced by heterogeneous forces at work that composed and connected its milieu of events and eventualities. In order to adequately consider the eruption of Calbuco, one would have to not only consider the history of events (involving both 'natural' and 'social' elements that compose

Beyond the Nature-Artifice Divide

the history of the event), but also the eventualities that mapped out and produced those events.

Deleuze and Guattari avail themselves of the concept of the rhizome in order to discuss how heterogeneous forces are connected, how 'chains of every nature are connected to diverse modes of coding (biological, political, economic, etc.) that bring into play not only different regimes of signs but also states of things of differing status'.[82] Literally, the rhizome is an underground stem that sprouts adventitious roots and lateral shoots, as is seen with ginger and turmeric. In Deleuze and Guattari's appraisal, rhizomatic emergence is opposed to the arborescent model of growth that postulates a beginning or *arche* in the form of a seed that follows a certain structure through a trunk and into a leafy tree top:

> Unlike trees or their roots, the rhizome connects any point to any other point, and its traits are not necessarily linked to traits of the same nature; it brings into play very different regimes of signs, and even nonsign states ... it has neither beginning nor end, but always a middle (*milieu*) from which it grows and which it overspills ... unlike a structure, which is defined by a set of points and positions, with binary relations between the points and biunivocal relationships between the positions, the rhizome is made only of lines ... the rhizome is an acentered, nonhierarchical, nonsignifying system ... that is totally different from the arborescent relation: all manner of 'becomings'.[83]

The middle or milieu is the essential dimension of the rhizome, the fact that it has neither beginning nor end, neither *arche* nor *telos*. The concept of the rhizome thus has an affinity with the concepts of difference in-itself and the concept of becoming. Whereas the Aristotelian classificatory schema attempted to capture difference through identities, the rhizome attempts to make connections between heterogeneous series without totalising them or classifying them within an arborescent structure of *arche* and *telos*, origin and progression. The rhizome designates the *connectivity* of heterogeneous series that disrupts the categories and kinds that define them disciplinarily – the rhizome designates only risings and fallings, passages and becomings.

The wasp and the orchid form an interesting pair with regard to the rhizome. Discussing Rémy Chauvin in *The Machinic Unconscious*, Guattari tells us:

> It is known that the wasp, effectuating a simulated sexual act with a morphological and olfactory lure constituted by the rostellum of the orchid,

> afterwards releases and attaches the pollen that it transports onto other plants, thus ensuring the cross reproduction of this species. The ensemble of the transcoding systems authorizing these round-trip tickets between the vegetable kingdom and the animal kingdom appears completely closed to any individual experimentation, training, or innovation ... Under these conditions, nothing would be gained by reducing a symbiosis like that of the wasp and the orchid to a simple 'attachment' between two heterogeneous worlds.[84]

The heterogeneous series that constitutes the symbiotic eventualities between the wasp and the orchid cannot be merely an imitation of the wasp by the orchid, the orchid 'playing wasp'. It is, rather, a question of becoming, the intensive delirium of both plant and insect crossing over and forming a series that conditions the event of pollination. Deleuze and Guattari discuss the process as a 'becoming-wasp of the orchid and a becoming-orchid of the wasp ... the two becomings interlink and form relays in a circulation of intensities ... Rémy Chauvin expresses it well: "the *aparallel evolution* of two beings that have absolutely nothing to do with each other".'[85] There is no question of correspondence between the wasp and the orchid – they do not communicate with one another (what could they say to one another?), but there are signs and triggers that provoke a series of events, thresholds crossed that constitute their attraction. I mention all of this because while the event of a wasp twitching within an orchid is an extensive phenomenon that one can date and classify, the eventuality of becoming-wasp and becoming-orchid that maps out the event is an immense problem stretching across ages and geological epochs, across an entire ecology of earth and organism. In order to adequately consider the emergence of the sensory receptivity implied in the encounter between wasp and orchid one requires more than the identification of wasps and orchids, plants and insects – one requires an idea of becoming, the eventuality of a differential production of heterogeneous series of events. In *Difference and Repetition,* Deleuze provides an illuminating example of the evolution of the animal eye as the production of two heterogeneous series:

> An animal forms an eye for itself by causing scattered and diffuse luminous excitations to be reproduced on a privileged surface of its body. The eye binds light, it is itself a bound light ... there is indeed an activity of reproduction which takes as its object the difference to be bound; but there is more profoundly a passion of repetition, from which emerges a new difference (the formed eye or the seeing subject).[86]

Here, larval syntheses on the surface of a body arrest luminous excitations, but this power to absorb, this production of passivity, at once produces another series – a completely different, heterogeneous, series – a seeing subject. The passivity of luminous reception thus produces the affection of a seeing subject. The forked event of an eye forming on the surface of the animal body – the passive receptivity and the affection of a seeing subject – is facilitated by an intensive becoming, or what Deleuze calls the repetition from which emerges a new difference.

With these examples in mind, we may now make some general comments about events and eventualities. Events are the side of time that is axiomatic, linear, discrete, horizontal and extensive. Eventualities are the side of time that is problematic, non-linear, continuous, vertical and intensive. I refer to them continually in the multiple because they would be abstractly and egregiously considered in the singular – time is always a multiplicity. But the most difficult problem is not what events and eventualities are but, rather, how to think of space as a product of time, how to think of extensivity as thoroughly embedded within intensivity. This is what the dual notions of events and eventualities tries to accomplish: to view spatial configurations as instants of time, actualised events, and to view the becoming within which those instants of time have their moment, within which they are composed and actualised, as virtual eventualities. Thus events are moments or states of eventualities, the present is the state of space, the past is the moment of memory and history, and the future is the moment of creativity, the new.

With regard to technology, we can see that events are *conducted* through technologies as so many states of affairs, but they are *produced* through heterogeneous syntheses, as the evolution of the eye produces a seeing subject, as the eye and the hand produce an anterior field, or, to borrow John Protevi's example, as lightning is 'produced from a field of electrical potential differences between cloud and ground'.[87] But differing potentials are of a different temporality, not quite on the order of events, but rather on the order of eventualities, which belong to a productive regime of flows, breaks and interruptions through the syntheses of machines. While the labourer can be represented as working for the event of getting paid, the real process of production must account for the entire investment upon the full body of capital, and the event of getting paid is merely a technical abstraction from the vast syntheses of eventualities that condition the event. While events can be scheduled on a calendar, the relative consequence between the eye and the hand that conditions the sense and investment of calendar time cannot be understood according to the same calendar. Technologies such as calendars are conscious mechanisms oriented toward scheduling, but the machines that produce those technologies are unconscious syntheses that span across an immemorial history. The conscious evaluation of time itself, in other words, is a product of an unconscious process that cannot be evaluated in the same terms. In this sense, there can be no priority of machines because the succession that such a priority would entail operates on the level of technology and events. Machines are not lower level conditions for technology, and neither are events subsequent to eventualities. Thus, the practical rationalities of techniques are not produced by machinic investments like a cause produces an effect; rather, production and product are identified in machinic process, so the events produced are produced by their investment *within* the immense syntheses of eventualities that constitute them. Events are products, but they are products that are immediately recast as productive elements within the process of eventualities that produced them. It is in this sense that practical rationalities are produced by machinic investments, although to be sure they fall back upon their investments, as technologies conduct events by intervening upon the eventualities that produce them. Thus, events and eventualities are both dimensions or regimes of the same temporal process of production.

The concepts of technology, milieu and machine are efforts to reinstall the human within her concrete environment: technically, in an environment composed of sense-events and effect-events; and machinically, in the milieus that compose that environment as so many eventualities acting transversally across distances of time and space. Eventualities are intensive, and the becomings or relations of intensities require a new metaphysics not based on substantial identities or relations of cause and effect. Thus Deleuze and Guattari avail themselves of terms like 'plane of consistency' and 'abstract machine' to designate how concrete machinic assemblages operate across distances of strata by constructing continuums of intensity.[88] But beyond the difficult jargon, the essential problem they are addressing is the problem of milieu, understood as the composition of events and eventualities. Foucault called *Anti-Oedipus* a book of ethics precisely because it encourages us to reconceptualise the composition of humans and environments, releasing desire from the constraining opposition between the human and nature. If we reconsider the *megalopsychon*'s deliberate step through the *agora* from the machinic perspective, we find that the *megalopsychon*'s affected movements, just like our own habits of movement, engage techniques of walking in the fashion of little Joey, embedded within a complex of machines that do not make a substantial distinction between the human and the non-human or the natural and the artificial: 'there is no such thing as either man or nature now, only a process that produces the one within the other and couples the machines together ... the self and the non-self, outside and inside, no longer have any meaning whatsoever'.[89] Deleuze and Guattari's concept of machine thus scrambles the lines that govern the hierarchical configuration of nature (ascending from the physical to the social, from the natural to the artificial) and, instead, draws the profile of a concept of nature along entirely productive lines, a composition of the world in asymmetrical and heterogeneous strokes, a world where a wasp is brought into delirious communication with an orchid, a new theory of milieu where nature is made entirely of art.

Notes

1. Gilles Deleuze, *Negotiations*, trans. Martin Joughin (New York: Columbia University Press, 1995), p. 155.
2. On this point, see Hadot, *The Veil of Isis*, pp. 91–8.
3. Philipp Descola provides a magisterial account of the distinction as the result of a particular ethnographic experience. Philippe Descola, *Beyond Nature and Culture*, trans. Janet Lloyd (Chicago: University of Chicago Press, 2013).
4. Michel Foucault, *Power*, ed. James D. Faubion and Paul Rabinow (New York: The New Press, 2000), p. 364; italics mine. For an analysis of the ethical dimensions of Foucault's philosophy of technology, see Steven Dorrestijn, 'Technical Mediation and Subjectivation: Tracing and Extending Foucault's Philosophy of Technology', *Philosophy & Technology*, Vol. 25, No. 2 (2012), pp. 221–41.
5. Foucault, *Power*, p. 364.
6. Aristotle, *Complete Works, Vol. 2*, pp. 1773–6. Further references to the *Nicomachean Ethics* are given in the main text.
7. In Book IV of the *Nicomachean Ethics*, Aristotle does not discuss the *megalopsychon* in terms of *tekhnē*, which is thematically treated in Book VI alongside *epistêmê* (knowledge). However, in Book IV Aristotle is concerned with a descriptive catalogue of virtues as they stand, while *tekhnē* designates the becoming or the practical production of contingent things. So while it is clear that *tekhnē* is not *arête*

(virtue), it is nonetheless involved in the *production* of *arête*. This highlights the essentially temporal dimension of *tekhnē* discussed below. Notably, Aristotle proceeds to discuss *megalopsychia* from the outset through its relation to external goods and objects: 'Greatness of Soul, as the word itself implies, seems to be related to great objects; let us first ascertain what sort of objects these are' (1123b1).

8. 'Space, Knowledge, and Power', in Foucault, *Power*, p. 364.
9. Bruno Bettelheim, 'Joey: A "Mechanical Boy"', *Scientific American*, March 1959.
10. Gilles Deleuze and Félix Guattari, *Anti-Oedipus: Capitalism and Schizophrenia*, trans. Robert Hurley, Mark Seem and Helen R. Lane (Minneapolis: University of Minnesota Press, 1983), pp. 4–5.
11. Michel Foucault, *Security, Territory, Population*, trans. Graham Burchell (New York: Palgrave Macmillan, 2007), p. 20.
12. Foucault, *Security, Territory, Population*, pp. 18–20.
13. Foucault, *Security, Territory, Population*, p. 19; italics mine.
14. Martin Heidegger, *Plato's Sophist*, trans. Richard Rojcewicz and André Schuwer (Bloomington: Indiana University Press, 1997), p. 29. In this regard, see also Bernard Stiegler, *Technics and Time, 1: The Fault of Epimetheus*, trans. Richard Beardsworth and George Collins (Stanford: Stanford University Press, 1998), p. 9.
15. Hippocrates, *On Airs, Waters, and Places*, trans. Francis Adams (Whitefish: Kessinger Publishing LLC, 2010), pp. 3–4.
16. Hippocrates, *On Airs, Waters, and Places*, pp. 4, 30.
17. Hippocrates writes, 'One ought also to be guarded about the rising of the stars, especially of the Dogstar, then of Arcturus, and then the setting of the Pleiades; for diseases are especially apt to prove critical in those days.' Hippocrates, *On Airs, Waters, and Places*, p. 11.
18. Hippocrates, *On Airs, Waters, and Places*, p. 23.
19. Foucault, *Security, Territory, Population*, pp. 20–1.
20. Foucault, *Security, Territory, Population*, pp. 20–1.
21. H. G. Alexander (ed.), *The Leibniz-Clarke Correspondence: Together with Extracts From Newton's Principia and Opticks* (Manchester: University of Manchester, 1956), p. 174.
22. Alexander, *The Leibniz-Clarke Correspondence*, p. 66.
23. Alexander, *The Leibniz-Clarke Correspondence*, p. 53; italics mine.
24. Noam Chomsky and Michel Foucault, *The Chomsky-Foucault Debate: On Human Nature* (New York: The New Press, 2006), p. 12. Chomsky repeats this point in several places. See his article, Noam Chomsky, 'Language and Nature', *Mind*, New Series, Vol. 104, No. 413 (January 1995), pp. 1–61; his book, *On Nature and Language* (Cambridge: Cambridge University Press, 2002); and his lecture, 'The Machine, the Ghost, and the Limits of Understanding: Newton's Contributions to the Study of Mind', at the University of Oslo, September 2011.

Chomsky, of course, thinks there is an absolute basis of human nature that grounds rationality, whereas Foucault does not separate rationality from the historical knowledge/power relations that facilitate it – a difference that produces their disagreement when the debate turns to politics and human nature. See Chomsky and Foucault, *The Chomsky-Foucault Debate*, p. 57.
25. Chomsky and Foucault, *The Chomsky-Foucault Debate*, p. 12.
26. Alan Gabbey, 'Newton, Active Powers, and the Mechanical Philosophy', in *The Cambridge Companion to Newton*, ed. I. Bernard Cohen and George E. Smith (Cambridge: Cambridge University Press, 2002), p. 341.
27. From a draft variant of Query 23 of the 1706 Latin edition of the *Opticks*, cited in Gabbey, 'Newton, Active Powers', p. 344.
28. Georges Canguilhem, 'The Living and Its Milieu', in *Knowledge of Life*, ed. Paola Marrati and Todd Meyers, trans. Stefanos Geroulanos and Daniela Ginsburg (New York: Fordham University Press, 2008), p. 99.
29. Canguilhem notes that Newton's *Optics* may be the first explanation of an organic phenomenon through the action of the milieu, albeit defined in terms of strictly physical properties. Canguilhem, *Knowledge of Life*, p. 100.
30. Canguilhem, *Knowledge of Life*, p. 100.
31. See Eric Schliesser, 'Newton's Substance Monism, Distant Action, and the Nature of Newton's Empiricism: Discussion of H. Kochiras "Gravity and Newton's Substance Counting Problem"', in *Studies in History and Philosophy of Science, Part A*, Vol. 42, No. 1 (2011), pp. 160–6. The concept of action at a distance in Newton has recently been the focus of lively debate. See also Hylarie Kochiras, 'Gravity's cause and substance counting: contextualizing the problems', in *Studies in History and Philosophy of Science, Part A*, Vol. 42, No. 1 (2011), pp. 167–84; and Andrew Janiak, 'Three Concepts of Causation in Newton', in *Studies in History and Philosophy of Science, Part A*, Vol. 44, No. 3 (2013), pp. 396–407. For an account of Newton's influence on natural philosophy throughout eighteenth-century England, see Robert Schofield, *Mechanism and Materialism: British Natural Philosophy in An Age of Reason* (Princeton: Princeton University Press, 1970).
32. Canguilhem, *Knowledge of Life*, p. 100.
33. Citing Comte's fortieth lesson of the *Course of Positive Philosophy*, Canguilhem, *Knowledge of Life*, p. 101.
34. Canguilhem, *Knowledge of Life*, p. 102.
35. 'If Comte anticipates the idea of a subordination of the mechanical to the vital – the idea he would later formulate in mythical form in *The System of Positive Polity* and *The Subjective Synthesis* – here [in the

Course of Positive Philosophy] he nevertheless deliberately represses it.' Canguilhem, *Knowledge of Life*, p. 103.
36. Canguilhem, *Knowledge of Life*, p. 103.
37. Canguilhem, *Knowledge of Life*, p. 103.
38. From Louis Roule's *La vie des rivières*, cited in Canguilhem, *Knowledge of Life*, p. 103.
39. Canguilhem, *Knowledge of Life*, p. 117.
40. Canguilhem, *Knowledge of Life*, p. 104.
41. Canguilhem, *Knowledge of Life*, p. 111.
42. Canguilhem suggests a geographical model to examine the relations between complexes: 'Geography has to do with complexes – complexes of elements whose actions mutually limit each other and in which the effects of causes become causes in turn, modifying the causes that gave rise to them.' Trade winds are exemplary in this context because they 'displace surface water that has been heated by contact with the air; the cold deep waters rise to the surface and cool the atmosphere; low temperatures engender lower pressure, which generates winds; the cycle is closed and begins again'. Canguilhem, *Knowledge of Life*, p. 109.
43. Jakob von Uexküll, *A Foray into the Worlds of Animals and Humans with A Theory of Meaning*, trans. Joseph D. O'Neil (Minneapolis: Minnesota University Press, 2010), pp. 44–52.
44. Von Uexküll, *A Foray into the Worlds of Animals and Humans*, p. 53.
45. Von Uexküll, *A Foray into the Worlds of Animals and Humans*, p. 52.
46. Von Uexküll, *A Foray into the Worlds of Animals and Humans*, p. 48.
47. Bruno Latour makes this point in his 2013 Gifford Lectures, 'Facing Gaia: A New Enquiry into Natural Religion'.
48. André Leroi-Gourhan, *Gesture and Speech* (Cambridge, MA: MIT Press, 1993), p. 35.
49. Gregory of Nyssa cited in Leroi-Gourhan, *Gesture and Speech*, p. 35.
50. Gregory of Nyssa cited in Leroi-Gourhan, *Gesture and Speech*, p. 35.
51. This reiterates Leroi-Gourhan's denial of a cerebral cause to bipedalism: the brain is incidental in Leroi-Gourhan, since it advances only as the face shortens and the mechanical stresses of the mandible diminish (since the forehead requires a rearrangement of the stresses of the anterior territories so that the occipital cavity can extend in the measure that the base of the skull or the mandible is shortened). Leroi-Gourhan, *Gesture and Speech*, pp. 71–5.
52. Below we discuss Deleuze's treatment of the evolution of the eye, where luminous excitations are bound, and a novel and asymmetrical difference is produced, namely the formed eye or the seeing subject. Deleuze, *Difference and Repetition*, p. 96.
53. Leroi-Gourhan, *Gesture and Speech*, p. 90.
54. Leroi-Gourhan, *Gesture and Speech*, p. 228.

55. Deleuze and Guattari claim that 'Leroi-Gourhan has gone the farthest toward a technological vitalism taking biological evolution in general as the model for technical evolution: a Universal Tendency, laden with all of the singularities and traits of expression, traverses technical and interior milieus that refract or differentiate it in accordance with the singularities and traits each of them retains, selects, draws together, causes to converge, invents.' Deleuze and Guattari, *A Thousand Plateaus*, p. 407.
56. Leroi-Gourhan finds, for example, that the transition from mythological thinking to rational thinking was a 'very gradual shift exactly synchronous with the development of urban concentrations and of metallurgy. The earliest beginnings of Mesopotamian writing date back to about 3500 B.C., some 2,500 years after the appearance of the first villages. Two thousand years later, toward 1500 B.C., the first consonantal alphabet appeared in Phoenicia . . . by 350 B.C. Greek philosophy was advancing by leaps and bounds.' Mythological or pre-rational thinking, on the other hand, deploys different (non-consonantal) techniques: 'The fact that verbal language is coordinated freely with graphic figurative representation is undoubtedly one of the reasons for this kind of thinking, whose organisation in space and time is different from ours and implies the thinking individual's continuing unity with the environment upon which his or her thought is exercised. Discontinuity begins to appear with agricultural sedentarisation and with early writing. The basis now is the creation of a cosmic image pivoted upon the city. The thinking of agricultural peoples is organised in both time and space from an initial point of reference – *omphalos* – round which the heavens gravitate and from which distances are ordered.' Leroi-Gourhan, *Gesture and Speech*, pp. 210–11.
57. Leroi-Gourhan, *Gesture and Speech*, pp. 97–183.
58. Gilles Deleuze, *Instincts et institutions*, ed. Georges Canguilhem (Paris: Hérissey à Évreux, 1971), p. xi. This text was originally published in 1953. Leroi-Gourhan presents his theory of the relation between the technological and the ethnic in *Milieu et techniques* (Paris: Albin Michel, 1945).
59. Deleuze and Guattari citing Bruno Bettelheim's *The Empty Fortress*, in *Anti-Oedipus*, p. 37.
60. Bettelheim, 'Joey: A "Mechanical Boy"', p. 3.
61. Deleuze and Guattari repeat this point throughout the two volumes of *Capitalism and Schizophrenia*. Discussing the production of metal weapons, they argue that the weapon material as well as its use is inseparable from a social formation. So while they identify a 'machinic phylum' or a 'technological lineage', 'at the limit, there is a single phylogenetic lineage, a single machinic phylum, ideally continuous: the flow of matter-movement, the flow of matter in continuous variation, conveying

singularities and traits of expression. This operative and expressive flow is as much artificial as natural: it is like the unity of human beings and Nature.' Deleuze and Guattari, *A Thousand Plateaus*, p. 406.
62. Daniel W. Smith, 'Deleuze and the Question of Desire: Toward an Immanent Theory of Ethics', in *Essays on Deleuze* (Edinburgh: Edinburgh University Press, 2012), pp. 182–3.
63. Deleuze and Guattari, *Anti-Oedipus*, p. 116.
64. Foucault, *Security, Territory, Population*, p. 10; italics mine.
65. Karl Marx, *Capital: A Critique of Political Economy, Volume One*, trans. Ben Fowkes (New York: Penguin, 1976), pp. 493–4.
66. Guattari notes that with '*techne*, there are ontogenetic elements, elements of the plan, of construction, social relationships which support these technologies, a stock of knowledge, economic relations and a whole series of interfaces onto which the technical object attaches itself'. Félix Guattari, 'On Machines', in Andrew Benjamin (ed.), *Complexity*, JPVA, No. 6 (1995), pp. 8–12. Similarly, Deleuze acknowledges that 'it's easy to set up a correspondence between any society and some kind of machine, which isn't to say that their machines determine different kinds of society but that they express the social forms capable of producing them and making use of them'. Deleuze, 'Postscript on Control Societies', in *Negotiations*, p. 180.
67. Deleuze and Guattari, *Anti-Oedipus*, pp. 4–5.
68. The mouth is not machinic in a substantial sense, it is itself composed of other machinic interrelations between teeth, saliva, tongue and lips, just like the larynx holds a machinic relation to thyroid glands, abducting and adducting muscles, and the epiglottis. The emphasis here is on inter-relational composition, which affords variable functionality.
69. Guattari, 'On Machines', pp. 8–12.
70. Gilbert Simondon, *On the Mode of Existence of Technical Objects* (Paris: Aubier, Editions Montaigne, 1958), p. 21. Lewis Mumford pursues a similar (although perhaps more rigid) position: 'the machine, if left to its own devices, goes in for standardization, mass production, automation'. Lewis Mumford, *The Urban Prospect* (New York: Harcourt, Brace & World, Inc., 1968), p. 7.
71. Gilles Deleuze and Claire Parnet, *Dialogues II* (New York: Columbia University Press, 2007), p. 69.
72. Deleuze and Guattari, *A Thousand Plateaus*, p. 313.
73. Michel Foucault, 'Theatrum Philosophicum', in *Aesthetics, Method, and Epistemology*, ed. James D. Faubion and Paul Rabinow (New York: The New Press, 1998), p. 349.
74. Charles Péguy, *Temporal and Eternal*, trans. Alexander Dru (Indianapolis: Liberty Fund, 2001), p. 4.
75. Foucault, 'Theatrum Philosophicum', p. 349; italics mine.
76. Alessandro Portelli, *The Death of Luigi Trastulli and Other Stories:*

Form and Meaning in Oral History (New York: SUNY Press, 1991), p. 1.
77. Portelli, *The Death of Luigi Trastulli*, p. 15.
78. Portelli, *The Death of Luigi Trastulli*, p. 21.
79. Portelli, *The Death of Luigi Trastulli*, p. 21.
80. Deleuze and Guattari, *A Thousand Plateaus*, p. 4.
81. Antonin Artaud, 'To Have Done with the Judgment of God', in *Selected Writings*, ed. Susan Sontag (Berkeley: University of California Press, 1976), p. 571.
82. Deleuze and Guattari, *A Thousand Plateaus*, p. 7.
83. Deleuze and Guattari, *A Thousand Plateaus*, p. 21.
84. Félix Guattari, *The Machinic Unconscious: Essays in Schizoanalysis* (Los Angeles: Semiotext(e), 2011), pp. 121–2.
85. Deleuze and Guattari, *A Thousand Plateaus*, p. 10.
86. Deleuze, *Difference and Repetition*, p. 96.
87. John Protevi, 'Deleuze and Life', in *The Cambridge Companion to Deleuze*, ed. Daniel W. Smith and Henry Somers-Hall (Cambridge: Cambridge University Press, 2012), p. 240. Cf. Deleuze, *Difference and Repetition*, pp. 118–19.
88. See Deleuze and Guattari, *A Thousand Plateaus*, pp. 70–3. Describing 'abstract machines', Guattari writes that they 'traverse various levels of reality and establish and demolish stratifications. Abstract machines cling not to a single universal time but to a trans-spatial and trans-temporal *plane of consistency*.' Guattari, *The Machinic Unconscious*, p. 11.
89. Deleuze and Guattari, *Anti-Oedipus*, p. 2.

Conclusion:
The Technological Composition of Milieus

A few years ago, I heard a story about an adolescent girl who, after moving from one psychiatric hospital to another, was admitted into the behavioural ward at a youth hospital in New Orleans. The behavioural ward is separated from the medical ward by a lawn bounded by a ruddy brick fence and a gate attended by a guard. She was admitted to the behavioural ward because a psychiatrist's evaluation had identified her condition as mental rather than medical. Some time later, she started screaming and shaking aggressively, exhibiting behaviour that is not uncommon in a psychiatric ward, and the hospital responded to it as they normally respond to such situations – which included a restraining protocol that kept her in a room for a period of time. Tragically, she died shortly afterwards. However, her autopsy revealed potassium levels that would have constituted a medical (and not a behavioural) emergency – an emergency that could have been easily treated had she just been transported across a small lawn, through a gate, and into an adjacent building. I mention all of this because this story of a person who grew up in psychiatric wards (and, lamentably, would eventually die in one) is possible only in a society that has profoundly separated the mind from the body. In other words, the conceptual architecture that separates the mind from the body is inscribed, as it were, within the institutional architecture that separates the behavioural ward from a medical ward. There is so much to negotiate, it turns out, between two adjacent buildings.

At the beginning of this book, we saw that the bifurcated concept of nature (split between primary and secondary qualities – a real objective nature over and against subjective appearances) was traced along the outlines of the modern knowing subject, a cogito in a machine. The bifurcation, like the cogito, was epistemologically configured: just as the modern subject (the embodied cogito) worries about its epistemological access to an objective world, so nature is split into primary qualities that can be known (i.e. that are mechanistically or scientifically intelligible) and secondary sensory qualities that cannot

Conclusion

be known (at least not in the same way). Now, what the above story illustrates is that one continues to see the bifurcation (along with the distinction between the mind and the body) everywhere – one can see it in the university system (that separates natural sciences from social sciences and humanities), and throughout the major institutions that characterise and govern human life. As previously mentioned, the problem with the distinction is that it places nature too far from the human, as something external to human institutions, practices and technologies. To be sure, any solution to the problem will likewise address the production of knowledge of nature as not a merely a conceptual issue but also as a concrete and practical problem. Thus, the main effort of this book has been, first, to identify the bifurcation of nature as an inadequate formulation of time and nature, and second, to offer conceptual resources that help not only to dismantle the bifurcation but also to move toward a new, more productive and concrete philosophy of nature based on time and events rather than space and substances. The concept of milieu, composed by technologies and machines, events and eventualities, is an attempt to sketch the profile of such a philosophy of nature.

Basically, there are two ways of considering entities, whether a creature, a hospital, a volcano, an orchid or a technical object. The first considers entities substantially, as something classifiable according to disciplinary categories amenable to the production of knowledge, while the other considers entities as events within a manifold of eventualities that are trans-disciplinary in nature, as their determining feature is not their identity but rather their differential constitution. A human life can be considered as a substantial mind-body composite moving through space, or as an arc of technically conducted extensive events taking place within a series of other events, which emerge from and transform an intensive manifold of eventualities that produced them. The former begins with the identification of a mind in distinction from the body, whereas the latter begins with differential processes that can account for the technical production of the distinction between the mind and body itself. The former upholds a linear causality wherein an agent is said to cause an event, whereas the latter reconsiders linear causality as an abstraction from the emergent complexity of a system, where heterogeneous eventualities transform the contours of events and the entities that they inform. Substances are identifiable, but events and eventualities take place in a differential and emergent register and therefore resist disciplinary categorisation (since events and eventualities both imply heterogeneous forces and elements that compose them, as the anterior field of perception implies the heterogeneous communication between hand and eye, as the eye itself implies a receptivity for light that at once affects a seeing subject, etc.). I do not mean to suggest that identities do not exist or that identities are somehow illusory – quite the opposite: identities are very real, but the question becomes how are they produced, through what technologies and machinic assemblages. Understood as an element of an event, an identity is no longer a substance but rather a result, a product. Furthermore, products are never final – they are, rather, events that are immediately recast within the immense series of events taking place in the process that produced them. Thus, we find that the identification of subjects and objects is a result of a procedure of abstraction from a fundamentally temporal process of events and eventualities. It is, of course, much easier to define a human (or an orchid, etc.) when taken in abstraction from the non-human ecology (or the wasp, etc.). But we realise that, while it is surely productive of disciplinary knowledge, the identification of entities through a procedure of abstraction from a concrete ecology of heterogeneous events should also be reevaluated according to a philosophy of nature whose object is precisely the wider ecology of events and eventualities that facilitate the production of disciplinary knowledge.

If we reexamine the bifurcation from this productive and compositional perspective, we find that the distinction between subject and object is not merely a disciplinary or conceptual distinction. The above story about a patient trapped, as it were, between the bifurcated architecture of a hospital is sufficient indication that the bifurcation is literally inscribed within the institutions and practices that compose human life. Moreover, we find that the bifurcation is better understood when seen in its practical habitat, which includes heterogeneous institutions, practices and technologies. This is not to say that the bifurcation informs our institutions and practices; there is not, first, bifurcation and then, subsequently, a formation of institutions, practices and disciplines. As we have seen, the production of knowledge is itself a thoroughly technical endeavour. Furthermore, as the parallel evolution of technics and milieu shows, forms of reasoning are traced upon forms of technologies. Thus, we now realise that an alternative (non-bifurcated) concept of nature would be accompanied by different technologies and machinic assemblages, different events and eventualities. In other words, while philosophy can draw the profile of a new concept of nature beyond

Conclusion

bifurcation, it will remain difficult to think about nature in terms of becoming without new practices, institutions and technologies – in short, until a new politics of nature emerges, which will transform not only our approach and understanding to nature, but also the ecology of events and eventualities that characterise the human/nature assemblage. In other words, we require a series of technological and machinic transformations in order for our knowledge of nature (along with its forms of production) to be reevaluated.

As previously mentioned, Nietzsche, Whitehead and Bergson are fundamentally in agreement about the fallacy of misplaced concreteness and the bifurcated concept of nature that emerges from it (which is oriented toward discursive knowledge, configuring nature in terms of space and matter, overlooking the fact that the fundamental datum of nature is time and becoming). They are also all in agreement that it remains difficult to think of nature in terms of time and becoming. Nietzsche described the fallacy as a function of 'forgetting' that was constitutive of discursive knowledge itself – a reification of entities that are fundamentally involved in a nexus of forces of becoming; and Bergson attributed the fallacy to an intellectual spatialisation of things, the intellect's tendency to think of things in terms of what we have called clock time (time understood in terms of space). Somewhat more positively, Whitehead claimed that it was not so much an error endemic to thought as such, but rather the 'accidental error of mistaking the abstract for the concrete'.[1] However, we are now in a position to offer another diagnosis of the fallacy. It is indeed difficult to think of nature in terms of events and eventualities, but not simply because it is a difficult exercise for the intellect: again, the intellect, like the production of knowledge, operates along technical lines. Disciplines technically carve out their fields from nature but are largely unable to reintegrate them, as we separate ourselves from something in order to know it – after all, something needs to become an 'object' in order to become an object of knowledge. We have already seen the technical nature of the separation between human and non-human, a separation that composed nature in the figure of the clock and its corresponding conception of mechanism. But now we realise that knowledge of nature may finally arrive at its own critique: we are thoroughly involved in what we are supposedly studying in a detached manner, there is no human/nature distinction, there is only a milieu of events and eventualities. In other words, if the condition of knowledge is a separation between the subject and the object (or between human and nature), then we must critically reevaluate the enterprise of knowledge itself. Furthermore, as Foucault shows so persuasively, the production of knowledge has always shared a trajectory alongside technology (as the principle of separating a subject from an object is at once a technological principle).[2] Thus, while it is difficult to conceive of nature as becoming, this is mainly due to the fact that our technicity does not support the thought – our technical evolution facilitates a different kind of knowledge, a different kind of disciplinary apparatus, a different kind of social order, a different kind of history and memory and a different conception of time and nature.

As Latour repeatedly emphasises, the ecological crisis thoroughly presses the problem of knowledge and nature – it is no longer possible to maintain that we are separate from a supposed nature that has been the object of the natural sciences.[3] The thin layer of earth within which our eventualities unfold now reasserts itself over and against the forms of reasoning that separate a subject from an object, the human from nature. The human/nature assemblage is undergoing a sea change whose eventuality, while unknown, will certainly challenge the ways that knowledge is produced. Additionally, there is ample evidence that technology is approaching a threshold of major significance for the evolution of the human. As we saw, the evolution of *Homo sapiens*, unlike other animals, has been in the direction of de-specialisation. Leroi-Gourhan portends that,

> Freed from tools, gestures, muscles, from programming actions, from memory, freed from imagination by the perfection of broadcasting media, freed from the animal world, the plant world, from cold, from microbes, from the unknown world of mountains and seas, zoological Homo sapiens is probably nearing the end of his career.[4]

But there is another interesting consideration in this line of thought. While our technical evolution may be characterised, thus far, as a series of externalisations of technologies (the shoe as a prosthetic foot, the oven as a prosthetic stomach, etc.), we are now witnessing the cusp of an age where our technologies (particularly biotechnologies) are becoming internalised. We now enjoy prosthetic joints and organs, and we can safely assume that our wearable technology will soon not only report internal biological data to doctors and other monitoring agents but also take on the role of calibrating certain biological functions. Thus, we already see technologies moving into the anatomical body, as it were. Furthermore, we realise that new social phenomena, new forms of thought and experience, will accompany

Conclusion

the 'bionic' human that is coming into focus. Nonetheless, the incorporation of such technologies is only a part of a more general trajectory of technology. As previously mentioned, the evolution of the human (that is, the anatomical, biological and physiological human body) has always been technically conducted (e.g. the recession of the mandible and the protrusion of the forehead that allowed for memory and speech functions were the result of bipedalism – an anatomical technicity – and the emergence of diverse kinds of tools and technical objects facilitated diverse social orders and practical rationalities; thus, it was not 'intelligence' or 'reason' that guided our bipedal evolution – quite the opposite). What is interesting now is that technology has arrived at a point where it is capable of designing and engineering the 'organic' element of human evolution. Now, the last two chapters of this book were, in a way, critiques of disciplinary categories such as the 'biological' and the 'social' in favour of categories such as technology and machine. Disciplinary categories are productive of knowledge, but this knowledge comes at a cost because one must abstract identities (as biological, anatomical, social, etc.) from a nexus of machinic connections or eventualities in order to isolate those disciplines in the first place. Furthermore, we find that the separation of disciplines (broadly in terms of the nature/artifice distinction but also specifically in terms of individual disciplines and sciences) is a technically conducted affair. Finally, the concept of machine attempted to cut across disciplinary categories in order to sketch a more concrete and eventual profile of human/nature assemblages. Thus, we may now raise the question of evolution in terms of technology and machines: what does our technology mean for evolution now, and what machines are at work as principles of our technology?

We are accustomed to discussing the current state of technology as inaugurating an era of connectivity, a world where everyone is connected and information is transmitted almost immediately. And we surely see new social phenomena, new affects and new forms of reasoning accompanying this era of technological connectivity (it is no coincidence that Arab Spring occurred in 2011 after mobile use reached critical mass in the relevant countries, and one does not need to look very far in order to see how technology is transforming our sense of and discussions about race, privacy, surveillance, gender, security, etc.). Now, while it is perhaps more difficult to notice, one can discern a certain kind of machinic principle operative throughout these diverse concerns, what we might call an *informational machine*. After all, it is not only nature that is understood in terms of information now – the human itself has also become informational. This should come as no surprise, since there is no reason to think that the technological form of reasoning that considers nature in terms of information would not consider the human in the same terms. But more profoundly, we note that this form of reasoning is merely a threshold in the unfolding of the technological enterprise that included the separation of the human from nature. In other words, the regime of knowledge that conceives of humans and nature in terms of information needed to go through a regime of knowledge that separates a subject from an object, the cogito from the clock. In order to produce knowledge of nature, the subject must be separated from it, but in order to render that knowledge informational, the subject and the natural object must configured in terms of clock time, the time of information, which can only then become the basic datum of knowledge of nature. Thus, we find that the procedure of abstraction that separated the human from nature, which is a constitutive procedure for isolating disciplinary categories (from the physical, chemical and the biological to the psychological, social and economic), also conditions the informational understanding of humans and nature. The basic point here is that the bifurcation of nature is preserved within the informational technology that now configures our regime of knowledge with its disciplinary apparatus, such that information is the basic unit of nature. However, as the world is redrawn in terms of information, the old disciplinary categories are undergoing disruption by the very forms of reasoning and technology that emerged from them. For example, we are at a point where we can genetically engineer human evolution, charting out a (quasi-natural) trajectory designed, at once biologically, but also technologically, i.e. through artifice. When technology, which has hitherto been 'externalised', begins to plot and design the 'internal' constitution of the human, there can be no question of nature/culture, or a subject-operator and an object-operated upon. So even with regard to human evolution, it is becoming less and less plausible to uphold the distinction between nature and artifice. In truth we have always been bionic in the sense that human evolution has paralleled technical evolution. But the informational human-nature-society assemblage, itself a manifestation of a technological principle that first separated us from nature, is now producing a regime of knowledge that may very well transform the bifurcation, not by a critique of knowledge, but rather through its culmination:

Conclusion

an informational machinic assemblage where the human is indistinguishable from artifice and nature becomes purely informational, data to respond to and manage. Nature would no longer be the great outdoors but rather a complex of data points to be factored through technologies engineered precisely in order to manage that complexity.

When considering a hopeful possibility for the future of the human, Leroi-Gourhan imagined 'the human of the near future as being determined by a new awareness and the will to remains *sapiens*':

> In such an event the problem of the individual's relationship with society will have to be completely rethought: We must face up squarely to the question of our numerical density and our relations with the animal and plant worlds; we must stop miming the behaviour of a microbic culture and come to grips with the management of our planet in terms other than those of a game of chance.[5]

It is clear that the future portends some ominous possibilities. We are faced with a real danger of destroying the conditions for our own possibility. But if the foregoing is correct, then we must disagree with Leroi-Gourhan's faith in knowledge or *sapientia*. If we are right, then it is precisely by remaining merely *sapiens* that we will arrive at a point where we are no longer recognisable as *Homo sapiens*. In that case, it is not the *sapiens* that will change, but rather the *Homo*. The technological enterprise of knowledge, purchased by the separation of the human from nature, would indeed transform our awareness, but only by transforming the milieu within which our destiny is composed. The evaluation of this new milieu, however, is emergent, and depends upon the complexities of its emergent composition. The will to knowledge is not so much human as it is technological, and the emergent milieu composed through technologies, whose production is at work within the intensive manifold of eventualities, will retrace or dissolve entirely the contours of the distinction between human and nature.

Notes

1. Whitehead, *Science and the Modern World*, pp. 50–1.
2. See the aforementioned technologies of exclusion, quarantine and epidemics and the subjects and objects of knowledge those technologies facilitated, in Foucault, *Security, Territory, Population*, p. 10.

3. See Bruno Latour's 2012–2013 Gifford Lectures, 'Facing Gaia: A New Enquiry into Natural Religion'.
4. Leroi-Gourhan, *Gesture and Speech*, p. 407.
5. Leroi-Gourhan, *Gesture and Speech*, p. 408.

Index

abstract, 2, 8–9, 10–14, 24, 34–5, 43, 48, 51–2, 69, 70–1, 73–4, 84–5, 95, 101, 109, 111, 130, 136, 144, 150, 153–4, 161, 163–5, 167–8
accident, 13, 95, 109, 119, 147, 165
activity, 33, 36, 40–1, 49, 53–4, 59, 112, 118–19, 128, 137, 146, 148, 152
actual, 54, 57, 63–4, 66–7, 75, 77, 85, 90
adequation, 94, 96, 114
affect, affection, 40, 47, 51, 66–7, 119, 124–5, 131, 145, 153, 155, 164, 167
animals, 30, 32, 45, 132, 134–5, 166
anthropocene, 7
appearance, 1–4, 8–9, 11, 14, 23, 51, 84, 93–4, 102, 105, 109, 113, 126, 159, 162
arche, 98, 103, 106, 151
Aristotle, 22, 27, 36, 48, 71, 86–9, 90–1, 95, 101, 107–8, 118, 119, 121–3
art, 117, 119, 121, 123, 148, 155
artifice, 117–19, 120, 125, 137–9, 167–9
attribute, 11–12, 26, 28, 40, 44, 55, 87, 109, 135, 165

becoming, 2–3, 14–15, 83, 85, 93, 95, 102–3, 107, 111, 117, 119, 121–3, 128, 133, 139, 146–7, 149, 151–4, 156, 165–6, 168
behaviour, 162, 169
belief, 4, 5–6, 16, 23, 27, 143
Bergson, Henri, 11–15, 17, 49, 50–6, 58–9, 60–9, 70–5, 83–5, 111, 165

bifurcation (of nature), 6, 7, 11, 12, 14, 21, 25–7, 29, 32, 41–3, 46–9, 62, 65, 68–9, 83, 85, 92–3, 118, 120, 162–5, 168
biology, 5–6, 77, 117, 125, 128–9, 130, 132, 146
body, 13, 15, 16, 22–4, 26–9, 30, 32, 42, 49, 50–5, 58–9, 60, 64, 66–7, 74, 76, 92, 119, 124–7, 129, 132–7, 149, 150, 152–3, 162–3, 166–7
brain, 13, 23, 28, 30, 42, 44, 52–4, 59, 60–2, 133–5, 158

categories, 34, 83–4, 86–9, 91–2, 95, 101, 107, 128, 130, 131–2, 144–5, 151, 163, 167–8
causation, 9, 36–9, 83
cause, 29, 36, 38–9, 40, 94, 134, 144, 154, 157–8, 163
certainty, 21–6, 41–2, 46, 56, 120–2
chemistry, 5–6, 117, 132, 146
clock, 27, 31, 32–9, 40–5, 47, 67–9, 118, 147, 149, 165, 168
cogito, 22–7, 31–4, 36–9, 40–2, 47, 68, 127, 150, 162, 168
Comte, August, 125, 128–9, 130, 132, 157–8
concept, 3–4, 7–8, 10, 14–15, 21, 23, 30, 33–6, 38, 40, 42–3, 48, 58, 67, 69, 71, 73–4, 83–9, 90, 92, 94–6, 98, 100–6, 109, 111–12, 114–15, 117–19, 120, 122–3, 125–9, 130–2, 137–9, 140, 142, 144–6, 150–1, 154–5, 157, 162–7
concrete, 9, 11–14, 22, 32, 48, 56–7, 60, 71–3, 95, 100–1, 107, 109, 111, 119, 140, 145, 149, 154, 163–5, 167

conduct, 21, 63, 98, 118, 120–3, 125, 132–3, 138–9, 153–4, 163, 167
conscious, 14, 27, 29, 32, 40, 48–9, 50, 52, 54, 58–9, 60, 66–9, 71, 118, 121, 136–9, 140, 143, 149, 154
content, 12, 69, 74, 78, 108, 110
contest, 4, 49, 91, 96–9, 100–1, 114
count, 24–5, 40, 43, 67

Deleuze, Gilles, 15, 21, 40, 43, 45–6, 49, 55, 60, 63, 69, 70, 76–9, 83–6, 89, 90–3, 98–9, 100, 110–13, 115–17, 120, 125, 135, 137, 139, 140, 142, 143–7, 149, 150–6, 158–9, 160–1
Descartes, Rene, 21–9, 30–2, 34, 36–9, 40–6, 48, 50, 54, 56, 68, 125–7, 131, 150
determination, 87, 90, 123, 125, 149
determined, 40–1, 50, 60, 86–8, 91, 96, 102, 106, 109, 122, 138, 149, 169
dialectic, 90–2, 107–9
difference, 14–15, 31, 33–4, 53, 55, 59, 62, 65, 66, 67, 68, 70, 74–5, 78–9, 83–9, 90–8, 99, 101–5, 107–9, 110–16, 131, 144, 146, 151–3, 157, 159
differences in kind, 70, 74, 84–5
differences of degree, 53, 70, 74, 84–5
discipline, 119, 120, 138, 164, 167
doubt, 22, 23, 37, 43, 64, 93, 101, 104, 109, 149, 159
duration, 39, 40–1, 43, 47, 60, 63, 65, 67–9, 71, 74–5, 78, 83–4
dynamic, 7, 36, 79, 99, 108

ecology, 42, 149, 152, 164–5
Eddington, Arthur, 1, 5–6, 9, 11, 12, 14–17
empirical, 2, 44, 83–4, 109, 110, 125–6, 130
equal, 35, 39, 64, 78, 94–5, 98–9, 102–4, 106, 126, 135
Eris, 97, 99
error, 13, 21–3, 26, 41–2, 47, 95, 148–9, 165

essence, 3, 39, 90, 93, 96, 102, 107, 109, 111, 113
events, 9, 14, 15, 16, 36, 47, 56, 62, 66, 69, 98, 111, 115, 117, 120, 121–5, 128, 132, 133, 135–6, 138–9, 140, 146–9, 150, 151–4, 163–5
eventuality, 69, 125, 128, 132–3, 137–9, 144, 146, 149, 150–4, 163–9
experience, 4–6, 9, 10, 12, 15, 26, 28–9, 57–8, 62, 66, 69, 74, 83–5, 94, 110, 155, 167
extension, 29, 30, 49, 65, 135–7
extensive, 30, 62, 65, 67–8, 71, 78, 146, 149, 150, 152–3, 163

fallacy of misplaced concreteness, 9, 11, 12, 13, 95, 165
filiation, 90, 105, 109, 137, 142
forgetting, 75, 95, 97, 108, 110, 165
form, 28, 33–4, 41–2, 46, 48, 53, 63, 71, 77–8, 84, 86, 88–9, 90, 96–7, 99, 100, 104, 108–9, 111, 112, 119, 133–4, 146, 148, 151, 158, 168

genus, 87–9, 90
Guattari, Felix, 15, 117, 120, 125, 139, 140, 142–7, 149, 150–2, 154–6, 159, 160–1

Heidegger, Martin, 3, 48, 121
history, 1–3, 5, 7–8, 11, 14–15, 21, 34, 36, 42, 44–6, 49, 57, 61, 72, 75, 91–4, 100, 106, 119, 120, 125, 146–9, 150–1, 153–4, 157, 161, 166
horizontal, 98, 100, 146–9, 153
human, 1–2, 6–9, 13, 15–16, 31–2, 34–6, 48–9, 58–9, 84, 88, 111, 115, 117–19, 120, 124, 132–9, 145–6, 154–5, 163–9

idea, 3, 5–7, 14–15, 21, 23–4, 30–1, 33–5, 38, 48, 67–8, 89, 90–2, 94–6, 99, 100–2, 104–9, 110–11, 115, 121, 152, 158

172

Index

idealism, 6, 21, 49, 50–1, 54–5, 69, 75, 84
identity, identities, 15, 25, 56, 83, 85–9, 94–5, 101–5, 108–9, 111–12, 151, 154, 163–4
illusion, 1–2, 5, 11, 14, 55, 59, 69
image, 3, 24, 42–3, 49, 50–9, 60–7, 76, 85–6, 91, 96, 99, 100, 108, 126, 144, 159
imagination, 23–5, 30, 43, 148, 166
immanent, 92–3, 100, 101, 105–7, 109
information, 10, 16, 36, 42, 167–9
intellect, 13–14, 17, 21, 75, 86, 95, 113–14, 127, 137, 165
intelligibility, 4, 27, 33, 35, 96, 108, 126–7, 137
intensive, 65, 67–9, 103–4, 152, 153–4, 163, 169

judgement, 4, 23–6, 37, 43, 92, 96–7, 102, 105, 132, 150

knowledge, 1, 2, 4, 6, 9, 10–12, 15, 16, 21–3, 26–7, 30, 42–3, 45–7, 51, 54–5, 59, 69, 73, 75, 78, 90, 95–6, 107, 117, 119, 121–2, 126, 128, 131–2, 137, 141, 145–6, 156–8, 160, 163–9

Lamarck, Jean-Baptiste, 125, 128–9, 130–1
language, 93, 96, 98, 113–14, 159
Leroi-Gourhan, André, 132–9, 141, 166, 169
life, 10, 22, 32, 34, 40, 52, 56–8, 68–9, 75, 97–8, 128, 131–3, 137, 142, 144, 147, 157–8, 161, 163–4
living, 32, 58, 66–7, 123, 131

machine, 14–15, 31, 32, 33, 35–6, 41, 47, 53, 59, 62, 117–18, 120, 125, 130, 139, 140, 141–6, 150, 153–5, 157, 160–3, 167–8
machinic, 52, 120, 139, 140, 142–5, 149, 151, 154–5, 160–1, 164–5, 167, 169

Marx, Karl, 111, 140, 142
materialism, 6, 21, 49, 75, 157
mathematics, 58, 73, 107–8
matter, 1, 4, 6, 11–17, 21, 29, 30, 33, 34–6, 39, 40–1, 48–9, 50–3, 55, 57, 58–9, 60, 62–3, 65, 69, 70, 76–8, 89, 101, 108, 111–12, 117, 127–8, 147, 160, 165
mechanics, 31, 51, 125
mechanism, 27, 30–6, 40, 44, 48, 52, 59, 95, 126, 132, 135, 154, 157, 165
mediation, 86, 88, 90–1, 101–3, 107
memory, 17, 30, 56–8, 60–3, 65, 69, 70, 76–8, 84–5, 109, 110, 112, 116, 136, 148, 153, 166, 167
mental, 1, 6–7, 11, 13–14, 25–9, 30, 32–3, 40–1, 48–9, 54, 59, 62, 69, 72–4, 95, 123, 127, 129, 130–2, 142
meson, 98
metaphor, 31, 66, 93–6, 142
milieu, 14–15, 60, 99, 117, 120, 123–5, 128–9, 130–3, 135, 137–9, 140–1, 145–6, 150–1, 154–5, 157, 159, 162–5, 169
mind, 1–3, 5, 8, 11, 13–16, 21, 23–4, 26, 29, 30, 32, 42–3, 49, 50–5, 62–3, 65, 76, 85, 91, 93–5, 97, 110–12, 117, 125, 127–8, 148, 153, 162–3
motion, 9, 27–8, 31, 33–4, 36–7, 41, 48, 57, 71, 108, 120, 126, 134, 140
movement, 3, 25, 33, 35–7, 49, 50, 52–3, 55–9, 60–9, 70, 72–5, 101, 103, 126, 130, 133, 137, 148, 155, 160
multiplicity, 68–9, 70–1, 73, 75, 78, 111, 145, 153

Newton, Isaac, 36, 48, 79, 125–9, 131
Nietzsche, Friedrich, 86, 92–9, 100–1, 103–4, 113–15, 122, 165
non-human, 6–8, 88, 111, 118–19, 120
number, 25, 43, 63, 71–3, 75, 78, 98, 116

173

object, objectivity, 1–3, 5–9, 10–15, 28–9, 31–2, 34, 37, 41–2, 45–7, 49, 50–1, 54–5, 59, 60, 63, 64, 66–7, 70, 74, 83, 86, 88–9, 92, 95, 101, 108, 110–11, 117, 119, 131, 133, 138, 140, 142, 143, 145, 147, 150, 152, 156, 160, 162–9
origin, 4, 13, 21, 37–8, 71, 90, 92–5, 97, 100, 103–5, 109, 110, 136, 140, 149, 151

passage of nature, 11, 13, 95
perception, 5, 12–14, 23, 30, 49, 50–9, 60–6, 76–7, 84–5, 164
physical, 1–2, 5–9, 10, 11–12, 14–16, 29, 32, 41–2, 44, 46, 49, 52, 55, 58, 63–5, 78, 108, 116–17, 119, 122, 125–8, 130–2, 139, 145, 147, 150, 155, 157, 168
physics, 1, 2, 5–6, 9, 12, 14, 34, 44, 46, 49, 56, 62, 69, 70–2, 78, 88–9, 90, 112–13, 117, 121, 128, 132, 146, 147, 154
plants, 32, 152
Plato, 2–3, 83, 86, 88–9, 90–6, 98–9, 100–9, 110–11
polis, 98
primary qualities, 3, 8, 27–9, 34, 47, 162
problem, 1–3, 6, 9, 12, 14, 21–3, 25–6, 41–2, 46–9, 59, 68, 74–5, 83, 86–9, 90–2, 97, 99, 100–3, 105–8, 110, 115, 120–1, 124, 128–9, 133, 141, 145, 148, 149, 152–4, 157, 163, 166, 169
process, 1, 5, 7–9, 10–12, 14, 28, 60, 63, 67–8, 70, 72, 77, 83, 95–6, 102–3, 111, 120–4, 136, 138–9, 142, 144–7, 152–5, 163–4
product, 24, 67, 94, 96, 112, 147, 153–4, 164
production, 1, 23, 36, 51, 84, 94, 111, 112, 120, 123, 138–9, 140–6, 149, 150, 152–4, 156, 159, 163–6, 169

qualities, 3, 8, 12, 14, 16, 21, 27–9, 30, 34, 41, 44–8, 50, 52, 67, 126, 145, 162

reality, 1–4, 6, 9, 10, 11, 12, 14, 40–2, 51, 64, 71, 73, 84, 96, 108, 120, 142, 143, 146, 147
reason, 21, 83, 87, 90, 94, 101, 112, 126, 137, 144, 157, 167
recognition, 38, 40, 57, 60–2, 76, 110
recollection, 60, 62, 110, 113
regime, 15, 127, 141, 146, 149, 151, 153–5, 168
reification, 95, 142, 165
representation, 13, 16, 25–6, 43, 47, 51, 53–5, 59, 60, 62, 64, 68–9, 70–1, 77, 83–8, 92, 101–5, 107, 111–13, 117, 130, 147, 149, 159
res extensa, 11–12, 27, 48, 127, 130
rivalry, 86, 97, 99, 102, 105–7

Sacks, Oliver, 56–7
science, 2–8, 13–14, 16, 21, 33–4, 38, 49, 55, 69, 70–1, 73, 114, 117, 119, 122, 127–8, 146, 157, 163, 166–7
secondary qualities, 3, 21, 27–9, 30, 34, 41, 44, 46, 48, 52, 162
selection, 58–9, 60–1, 63, 65, 88–9, 90–2, 133, 136, 148
sense, sensibility, 1–5, 8–9, 10–12, 16, 23–8, 30–9, 41, 44, 50–2, 54, 56–7, 60, 62–3, 65–6, 69, 73–4, 83–4, 87, 91–7, 99, 100–2, 104, 107–9, 110–11, 118–19, 120–2, 126–7, 129, 132–3, 135–8, 140, 143–4, 149, 150, 152, 154, 160, 162, 167–8
Simondon, Gilbert, 115, 143–4, 160
simultaneity, 39, 67, 75
solution, 71, 88, 92, 99, 102, 107–8, 163
space, 1, 5, 9, 11–14, 21, 27, 41–2, 48–9, 54, 62, 64–9, 70–1, 73–4, 77–8, 83–4, 98, 103, 106, 108, 114–15, 119, 120, 123–5, 127–9, 131, 133, 135, 139, 141, 145, 153–4, 156, 159, 163, 165
spatial, 11, 13–15, 41, 68, 72–5, 78, 83–5, 98, 123–4, 127, 129, 133, 135, 144, 153, 161, 165
spatialisation, 13–14, 165

Index

species, 87–9, 90, 95, 108–9, 130, 152
standardisation, 143–4
static, 79, 103
subject, subjectivity, 1–3, 5–9, 11, 14–16, 27, 29, 30, 32, 34, 40–2, 46, 48–9, 51, 54–5, 65–7, 83–4, 89, 92, 96, 111, 117, 131, 133, 140–1, 143–6, 149, 150, 152–3, 155, 158–9, 162, 165–6, 168, 169
substance, 11, 23–7, 29, 30, 32–4, 38, 40–2, 48, 66, 86, 103, 107, 117, 120, 126–7, 129, 145, 147, 150, 163–4
succession, 34, 37–9, 40–1, 67, 69, 75, 154

technical, 32, 34, 98, 111, 121, 133, 136–9, 140, 143, 147, 153–5, 159, 160, 163–8
technicity, 132–9, 149, 166–7
temporality, 39, 40, 139, 144, 149, 153
theory, 46, 52, 58, 63, 79, 92–4, 99, 105, 108, 115, 139, 145, 147, 155
transcendent, 31, 34, 83–4, 92, 99, 100, 105, 107, 109

transcendental, 79, 83–5, 104, 110
truth, 1–5, 11–12, 21, 24, 53, 92–4, 96, 100, 113, 117, 122, 168

umwelt, 132–3, 138
unconscious, 96, 120, 137–9, 140, 149, 151, 154, 161
understanding, 6, 25, 26, 28, 43, 45, 52, 67, 69, 74, 87, 108, 126, 165, 168
undetermined, 86–7

Vernant, Jean-Pierre, 97–9
vertical, 98, 100, 147, 149, 153
Vidal-Naquet, Pierre, 98, 106
virtual, 63–6, 72, 75, 77, 150, 153
von Uexküll, Jacob, 132–3

Whitehead, Alfred North, 3–4, 6, 9, 10, 11–17, 27, 46, 49, 78–9, 83, 95, 114, 165
world, 1–5, 7–8, 12–14, 21–7, 31, 34–9

Zeno's paradox, 70, 74